T0318593

The Arctic

The Arctic

A Barometer of Global Climate Variability

Neloy Khare

Rajni Khare

ELSEVIER

Elsevier
Radarweg 29, PO Box 211, 1000 AE Amsterdam, Netherlands
The Boulevard, Langford Lane, Kidlington, Oxford OX5 1GB, United Kingdom
50 Hampshire Street, 5th Floor, Cambridge, MA 02139, United States

Notices
Knowledge and best practice in this field are constantly changing. As new research and experience broaden our understanding, changes in research methods, professional practices, or medical treatment may become necessary.

Practitioners and researchers must always rely on their own experience and knowledge in evaluating and using any information, methods, compounds, or experiments described herein. In using such information or methods they should be mindful of their own safety and the safety of others, including parties for whom they have a professional responsibility.

To the fullest extent of the law, neither the Publisher nor the authors, contributors, or editors, assume any liability for any injury and/or damage to persons or property as a matter of products liability, negligence or otherwise, or from any use or operation of any methods, products, instructions, or ideas contained in the material herein.

Library of Congress Cataloging-in-Publication Data
A catalog record for this book is available from the Library of Congress

British Library Cataloguing-in-Publication Data
A catalogue record for this book is available from the British Library

ISBN: 978-0-12-823735-9

For information on all Elsevier publications visit our website
at https://www.elsevier.com/books-and-journals

Publisher: Candice Janco
Acquisitions Editor: Marisa LaFleur
Editorial Project Manager: Chris Hockaday
Production Project Manager: Kumar Anbazhagan
Cover Designer: Alan Studholme

Typeset by TNQ Technologies

Dedicated
to
Bharat Ratna
Late Dr. Avul Pakir Jainulabdeen Abdul Kalam

(15 October 1931—27 July 2015)

Science is a beautiful gift to humanity; we should not distort it.

The Pride of India, Dr. Avul Pakir Jainulabdeen Abdul Kalam, was a revered Indian aerospace scientist of international repute, who served as the 11th President of India from 2002 to 2007. He was born and raised in Rameswaram, Tamil Nadu (India), and studied physics and aerospace engineering. With an illustrious scientific career, he dedicatedly served four decades of his life as a Scientist and Science Administrator, mainly at the Defence Research and Development Organisation and Indian Space Research Organisation. He was intimately involved in India's civilian space program. His untiring and relentless efforts to strengthen and shape the country's military missile development initiatives through the development of ballistic missile and launch vehicle technology, which made India stand apart in its defence pursuit, gained him the glory of being the *Missile Man of India*. He played a pivotal role in India's *Pokhran-II* nuclear tests in 1998.

Dr. APJ Abdul Kalam, widely adored person and referred to as the "People's President," returned to his civilian life of education, writing, and public service after completing his presidential term. Apart from being a devoted scientist par excellence, he was an author, philosopher, philanthropist, a motivator, and above all, an embodiment of selflessness and simplicity.

He received several prestigious awards, including the Bharat Ratna, India's highest civilian honor. Unfortunately, while delivering a lecture at the Indian Institute of Management Shillong, Meghalaya (India), Dr. Kalam collapsed and died from an apparent cardiac arrest on July 27, 2015. His dream for motherland is well reflected in the following statement:

India has to be transformed into a developed nation, a prosperous nation and a healthy nation, with a value system.

APJ Abdul Kalam

Contents

16. Militarization of the Arctic

Foreword

Indubitably the "Arctic" region is known to be harsh but harbours a unique ecosystem and offers challenges that are central to the stability and sustainability of the planet earth. It can be counted as the last unspoiled pristine frontier. As the unprecedented global warming advances, its unforeseeable impacts over the Arctic region ought to be studied and understood in depth. Any subtle change in the ecosystem and the climatic conditions of this can have far-reaching global consequences.

The accelerated ongoing climate-linked changes compared to their geological antecedent, associated with the amplification mechanism, are fraught with unpredictable outcomes. This has added to the concerns of all climate scientists and other stakeholders about the emergent devastating scenario if the issues are not appropriately managed through adequate mitigation and adaptation strategies. The looming global risk involved due to the potential impact of climate could perhaps be minimized and controlled if projected futuristic trends are made available to help prepare humankind in a larger perspective.

India has been actively pursuing arctic research commensurate with its scientific strengths to enhance its global presence. However, the common man is still not fully aware of the multifaceted dimension of the Arctic and India's initiatives and contributions to Arctic research.

The present book titled *The Arctic: A Barometer of Global Climate Variability* has been conceived and crafted to provide a comprehensive overview on the Arctic, covering many important aspects; namely, physical structure, polar lights and midnight sun, Arctic geospace, Arctic explorations with historical perspectives, living and non-living resources, indigenous people of the Arctic, governance of Arctic affairs, climate change and its environment impact, militarization and other new opportunities the Arctic offers. Written in a reader-friendly format that is intrinsically informative and enjoyable, I am sure this book will be a good source of information about the Arctic region for the general public as well as students/professionals, and will help sensitize the wider population on the critical role that the region is destined to play in the context of global climate impacts and ecosystem sustainability.

(Eminent Indian Scientist)

Date: November 2020
Place: Hyderabad

Preface

The Arctic is not only known for polar bears and indigenous people, but also because it helps keep our world's climate in balance. Arctic Ocean ice acts as a large white reflector, bouncing some of the Sun's rays into space, and thus helping the world to maintain a good temperature. The recent anthropogenic impact has resulted in the warming up of the Arctic region over the past few decades—about twice as much as the global average. It has further magnified the impact because an increase in the sea ice melts means there is less to reflect the rays, and more heat is absorbed by the ocean. The Arctic regions are home to some of the most remarkable wildlife on Earth. They are also among the most vulnerable to the effects of climate change. We are gathering clear evidence of the regional impacts and global consequences of the changing climate in the polar regions.

The warming of the Arctic is prompting the movement of not only fish (such as cod) to farther north from previously cooler waters but also the Nordic countries' indigenous people, the Sami, to adopt changes to their way of life caused by milder temperatures as the winters are getting warmer and the vegetation is changing. Consequently, the reindeer that graze in winter cannot reach lichen. It implies **that the ecology of the Arctic** will change dramatically over the next decades. It is a matter of fact that such changes are occurring to the Arctic ecosystem, but we know very little about it.

Earth's poles (Arctic and Antarctic) are cold mainly because they get less direct sunlight than lower latitudes do. However, there is additionally another reason: ocean ice is white; thus, it reflects most daylight back to space. This reflectivity, known as "albedo," helps keep the poles cold by limiting their heat absorption. As shrinking ocean ice exposes additional water to daylight, the ocean absorbs additional heat that successively melts additional ice and curbs reflective power even more. This creates a feedback loop, one amongst many ways in which warming begets additional warming. Not like the Antarctic, the Arctic is an icy ocean encircled by land. (Antarctica is icy land surrounded by sea). The Arctic is home to around four million peoples. Instantly, the endemic communities have uniquely adapted themselves to the extreme environmental and climate conditions of the Arctic, which remains an enigma to the scientists.

Global warming is triggering changes in the polar environment. Temperatures in the Arctic region vary, from -60 to $+30°C$ (occasionally even hotter), and the sea ice during the summer months is declining. But with global warming, that ice is shrinking fast. Scientists predict there is also just about no summer ocean ice within the Arctic in the near future. We need to stop global warming, urgently. Temperatures in the Arctic are rising at twice the global rate, sparking an array of changes unlike anything seen in recorded history. Undoubtedly, the pace of global warming is staggering.

The Arctic additionally helps flow into the world's ocean currents, moving cold and heat water around the globe. Therefore, we value the Arctic as being very crucial for lots of reasons. We need help to tackle climate change, and to safeguard the Arctic from its worst effects as climate change is the greatest environmental challenge the world has ever faced. It is the need of the hour that masses are made aware about the facts of the Arctic and sensitize them regarding various aspects of this climatically sensitive region.

One of the alarming signals is that the Arctic ocean ice is currently declining by about 13% per decade, with the 12 lowest seasonal minimums all recorded in the last 12 years. Some climate model-based estimates suggest that between 2030 and 2050, the Arctic's sea ice will disappear completely during the summer months due to the recorded facts that not only the melting of ice sheets covering an area of about 70,000 square kilometers a year along with a sharp rise in temperatures since the end of the 1990s is taking place but also there is a failure of sea ice to recover ground lost during the summer months. It is easy to understand that the catastrophic changes in sea ice cover in the Pacific section of the Arctic Ocean can cascade very quickly with exponential ramifications.

It is widely that the main catalyst is human-induced climate change, boosted by a feedback loop known as Arctic amplification. The basic problem has become well known even among common people. While many people realize humans are indirectly undermining sea ice via global warming, it is well established that global warming boosts severe weather in general, but sea ice loss also favors bigger storms in the Arctic itself. Unbroken swaths of ocean ice usually limit what proportion of moisture moves from the ocean to the atmosphere, making it more durable for sturdy storms to develop. Due to the recent decline in the summer sea-ice extent, there is an expected increase in storms and waves, which could be responsible for coastal erosion.

Many Arctic people rely on seals and alternative native animals for food; however, the deterioration of ocean ice will make it progressively troublesome and dangerous to pursue certain prey. Hunters must not only wait longer for ice to form but also travel farther over mushier terrain. Farther offshore, the receding ice is commonly deemed excellent news for the oil, gas, and shipping industries. Less ocean ice additionally helps the ocean absorb additional carbon dioxide from the air, removing a minimum of several heat-trapping gases from the atmosphere. But as a consequence, it leads to the increasing acidity of the Arctic Ocean, as the waters of the ocean have become more acidic in line with increasing emissions of greenhouse gases. This phenomenon has dire consequences leading to the extinction of certain marine organisms, especially plankton, altering the ocean's entire ecosystem.

Observer states given to India mandates it necessary to forge favorable relations and form alliances with the coastal states within the present framework of Arctic governance. Tangible scientific efforts have been made in this direction including setting up an Indian Research Base in "Himadri." It is therefore imperative to assimilate, collate, and provide in an easy manner multidimensional aspects of the Arctic region. The region that witnesses the midnight sun, a natural phenomenon that occurs in the summer months in places north of the Arctic Circle or south of the Antarctic Circle, when the Sun remains visible at the local midnight, must be properly understood by common people.

Having realized the significance of bringing out detailed information about the scientifically significant naturally beautiful and geopolitically active region "The Arctic", this book titled *The Arctic: A Barometer of Global Climate Variability* is conceived and written, which covers all major issues like the Arctic's importance and physical structure, polar lights and midnight sun, geospace and arctic space weather, Arctic explorations with its historical perspectives, and the society and living style of its natives. The book also describes flora, fauna, and other natural resources of Arctic. Details of Arctic Ocean circulation highlighting unique features and significances are included in this book. It also projects the envisaged opportunities and potential of the Arctic Ocean in the near future. Besides, the governance of Arctic affairs is ably handled in this book encompassing International Arctic Research Initiatives vis-à-vis the Indian Arctic Exploration Programme to assess the impact of climate change on the Arctic as well as understanding the effect of the Arctic on global climate. In the recent past, the global competition for exploitation of the Arctic for economic gains has posed various environmental risks and necessitated militarization in Arctic. Both these important aspects are aptly addressed in this book.

We hope that this book will be a ready reference for scientific investigators in the field of Arctic Science and act as a catalyst for budding researchers to take up Arctic Science as a challenging career.

Neloy Khare
Rajni Khare

Date: November 2020
Place: New Delhi

Acknowledgments

The authors gratefully acknowledge the kind support and help received from many individuals working in the Arctic region. Polar Researchers like Prof. A.A. Mohamed Hatha from the Department of Marine Biology, Microbiology and Biochemistry, School of Marine Sciences, Cochin University of Science and Technology (CUSAT), Cochin, Kerala (India), Dr. Prashant Singh from Department of Botany, Institute of Science, Banaras Hindu University (BHU), Varanasi (India), Ms. Marika Marnela (Norway), and Dr. Shabnam Choudhary from the Ministry of Earth Sciences (MoES), New Delhi (India), are deeply admired and acknowledged for kindly providing their own photographs taken during their respective visits to the Arctic. The Arctic Portal, Arctic Council, Norwegian Polar Institute, International Boundaries Research Unit, Durham University, and their associated scientists are gratefully acknowledged for kindly permitting us to use their valuable photos/maps/data/materials in this book. Their generosity and kind helping attitude shall always be remembered by us. The materials/photos/images obtained from various websites are also duly acknowledged at appropriate places in the text of this book. All online contributors/photographers and weblink holders are sincerely and deeply acknowledged. We owe a lot to the providers of the abovementioned inputs in different forms based on which this book could be much improved, refined, and advanced in its presentation. We put on record our great honor, respect, and gratitude to the unknown contributors of freely available material/photographs/images on Wikipedia portal, which have given an added impetus to the preparation of this book in its present form.

Dr. Avinash Kumar and his group from the National Centre for Polar and Ocean Research (NCPOR), Goa (India), Dr. A.J. Luis and his group from NCPOR, Goa (India), Dr. Ratan Kar and his group from Birbal Sahni Institute of Palaeobotany, Lucknow (India), Prof. A.L. Ramanathan and his group from the Jawaharlal Nehru University, New Delhi (India), Dr. Sunil Sonbawne and his group from Indian Institute of Tropical Meteorology, Pune (India), Prof. A.K. Gwal and his group from Rabindranath Tagore University and Barkatullah University, Bhopal (India), Dr. Anu Gopinath and her group from Department of Aquatic Environment Management, Kerala University of Fisheries and Ocean Studies, Kochi, Kerala (India), Prof. A.A. Mohamed Hatha from CUSAT, Cochin (India), and Dr. Prashant Singh and his group from BHU, Varanasi (India), are gratefully acknowledged for sharing their scientific findings on the Arctic region as personal communication.

We express our sincere thanks to the MoES, Government of India, New Delhi (India), and the NCPOR, Goa (India), for various inputs, support, and encouragements. Author (NK) specially thank Dr. M.N. Rajeevan, Secretary Ministry of Earth Sciences, Government of India, for his kind support. Author (NK) continuously received guidance and encouragements from former Secretaries of MoES, namely Prof. V.K. Gaur, Dr. A.E. Muthunayagam, Dr. Harsh K. Gupta, Dr. Prem Shankar Goel, and Dr. Shailesh Nayak, and former Directors of NCPOR, namely Prof. Prem Chand Pandey, Dr. Rasik Ravindra, and Dr. S. Rajan. The present Director, NCPOR, Goa (India), Dr. M. Ravichandran, Dr. Rajiv Nigam, Former Chief Scientist, National Institute of Oceanography, Goa (India), Prof. Anil K. Gupta, Indian Institute of Technology, Kharagpur (India), Prof. Talat Ahmad, Vice Chancellor, Kashmir University, Jammu and Kashmir (India), Prof. Kalachand Sen, Director of Wadia Institute of Himalayan Geology, Dehradun (India), Dr. Vipin Chandra, Joint Secretary to Government of India at MoES, New Delhi (India), and Dr. K.J. Ramesh, Former Director General, India Meteorological Department, New Delhi (India), have always been a true well-wisher and supporter. Dr. Om Prakash Mishra of the National Centre for Seismology, New Delhi (India), and Prof. Rajesh Kumar Dubey, M.L. Sukhadia University, Jodhpur (India), are also acknowledged for providing many valuable suggestions at various stages of the preparation of this book.

The valuable help rendered by Prof. Avinash Chand Pandey, Director Inter University Accelerator Centre, New Delhi (India), is deeply acknowledged, which has become the starting point of the present book. Akshat Khare and Ashmit Khare have unconditionally supported us enormously during various stages of this book. Shri. Haridas Sharma from MoES, New Delhi (India), has helped immensely in formatting the text and figures of this book and bringing the book to its present form. Publishers (Elsevier) have done commendable job and are sincerely acknowledged.

Neloy Khare
Rajni Khare

Date: November 2020
Place: New Delhi

Abbreviations

AEC	Arctic Economic Council
AERC	Arctic Environment Research Centre
AERONET	Aerosol Robotic Network
AOD	Aerosol optical depth
ASCAT	Advanced Scatterometer
AWI	Alfred Wegener Institute
BHU	Banaras Hindu University
BSIP	Birbal Sahni Institute of Palaeobotany
BU	Barkatullah University
CAA	Chinese Arctic and Antarctic Administration
CCT	Climate Change Tower
CMEs	Coronal mass ejections
CMS-LS	Climate and Marine Science Life Cycle Assessment
CNR	National Research Council of Italy
CUSAT	Cochin University of Science and Technology
CWP	Current warm period
DIPN	Diisopropylnaphthalene
EEZ	Exclusive economic zone
EU	European Council
GGS	Global Geospace Science
GNSS	Global Navigation Satellite System
GPS	Global Positioning System
HSZ	Hydrate stability zone
IBRU	International Boundaries Research Unit, Durham University
IHO	International Hydrographic Organization
IIT	Indian Institute of Technology
IITM	Indian Institute of Tropical Meteorology
IMD	India Meteorological Department
IMO	International Maritime Organization
INTERACT	International Network for Terrestrial Research and Monitoring in the Arctic
IPEV	International Private Equity and Venture Capital Valuation
ISM	Indian summer monsoon
IUAC	Inter University Accelerator Centre
JNU	Jawaharlal Nehru University
KOPRI	Korea Polar Research Institute
KU	Kashmir University
KUFOS	Kerala University of Fisheries and Ocean Studies
LGM	Last Glacial Maximum

LIA	Little Ice Age
LIDAR	Light Detection and Ranging
LOM	Labile organic matter
MARPOL	International Convention for the Prevention of Pollution from Ships
MCA	Medieval Climate Anomaly
MLSU	M.L. Sukhadia University
MoES	Minister of Science and Technology and Earth Sciences
MoU	Memorandum of understanding
MTOE	Million tonnes oil equivalent
NASA	National Aeronautics and Space Administration
NATO	North Atlantic Treaty Organization
NCPOR	National Centre for Polar and Ocean Research
NCS	National Centre for Seismology
NDACC	Network for the Detection of Atmospheric Composition Change
NEP	Northeast Passage
NERC	Natural Environment Research Council
NGO	Nongovernmental organization
NILU	Norwegian Institute for Air Research
NIPR	National Institute of Polar Research
NIO	National Institute of Oceanography
NMA	Norwegian Mapping Authority
NPI	Norwegian Polar Institute
NSR	Northern Sea Route
NWP	Northwest Passage
NySMAC	Ny-Ålesund Science Managers Committee
OSCAT	Open Source Community for Automation Technology
PPs	Permanent participants
PRIC	Polar Research Institute of China
REE	Rare earth element
RTU	Rabindranath Tagore University
SAR	Specific absorption rate
SAT	Surface air temperatures
SIC	Sea ice conductivity
SIE	Sea ice extent
SIT	Sea ice temperature
SIV	Sea ice volume
SOLAS	International Convention for the Safety of Life at Sea
TEC	Total electron content
THC	Thermohaline circulation
TPP	Transpolar Passage
UiG	University of Groningen
UNCLOS	United Nations Convention on the Law of the Sea
USGS	United States Geological Survey
WIHG	Wadia Institute of Himalayan Geology

Chapter 1

Arctic—importance and physical structure

The moment we hear the word, "Arctic," a figure of ice-covered remote and barren land appears in our minds. For some people, Arctic is a synonym for the Arctic Ocean. But the Arctic region encompasses the northern parts of Asia, Europe, and North America in addition to the Arctic Ocean. The Arctic Ocean is located in the center of the northern part of Asia, Europe, and North America and within that is the North Pole. For some geographers, the Arctic is a mere extension of the Atlantic Ocean rather than an independent ocean. Thus, the North Pole is situated in the Arctic Ocean and the South Pole is situated on the Antarctic continent.

The word "Arctic" originates from the Greek word "*Arctikos*" which means "the land of Great Bear." Great Bear is the constellation of the northern hemisphere that revolves around the North Star. The Arctic region has been defined in various ways. Generally, the region located in the north of the Arctic Circle is considered as the Arctic. The Arctic Circle is an "imaginary line that circles the globe at approximately 66 degree 33′ N." It marks the latitude above which the sun does not set on the summer solstice and does not rise on the winter solstice [1].

This region comprises both land and ocean and is enclosed by Arctic Circle. It can also be defined as the region where the average temperature for the warmest month (July) is below 10°C. The northernmost "Tree Line" roughly follows the isotherm at the boundary of this region. An additional way to define the Arctic is where the 10°C summer isotherm is located. This is the line above which it is always colder than 10°C. The "Tree Line" is an imaginary line, and there are no forests to the north of this line. But in the Arctic region, various species of plants and animals are found to the north of the tree line. Some of the land parts of the Arctic are covered with ice sheets like Greenland; whereas, others have lush tundra like Alaska. These areas also inhabit large mammals like caribou, bears, wolves, and foxes, and a variety of plants. In summer times, migratory birds and other wildlife come to the Arctic to raise their young ones. Generally, the Arctic region comprises coastal regions on the north of the Arctic Ocean—Alaska, Canada, Norway, Sweden, Finland, and Russia (Siberia); entire Greenland and major parts of Iceland. Generally, Iceland is not considered part of the Arctic but the islands namely

The Arctic. https://doi.org/10.1016/B978-0-12-823735-9.00017-5

1

Svalbard, Franz Joseph land located in the Arctic Ocean are part of the Arctic Region. As Iceland is located to the south of the Arctic Circle, it is not considered as part of the Arctic.

The "tree line" and the 10°C isotherms play vital roles in defining the geography of the Arctic region. According to some Geographers, the 10°C isotherm line crosses through the southern part of Greenland, Goose bay of Canada, Labrador, Churchill, bays of Hudson and Liverpool and ends in Beaufort Ocean. After that turning toward the south, it crosses through Brooks's ranges of Alaska and moving forward with Baring Ocean it reaches the Aleutian group of Islands. From these Islands, it turns toward the north and reaches the Andy Bay of Siberia. Now it moves toward the north in Western Russia and crosses along the north coast of Norway and again turns to the south and reaches the Southern coast and makes a full circle. This encircled region is known as the real Arctic.

In the south of the Real Arctic is located the subarctic region. During the winters the region becomes as cold as the Arctic but during summer it is humid. Located in the North of 10°C isotherm line is the region where the average temperature for the whole year except 4 months is below 10°C. This region comprises the middle region of Siberia and Alaska and northern regions of Canada and Europe (Figs. 1.1A−C). Due to economic and social impact,

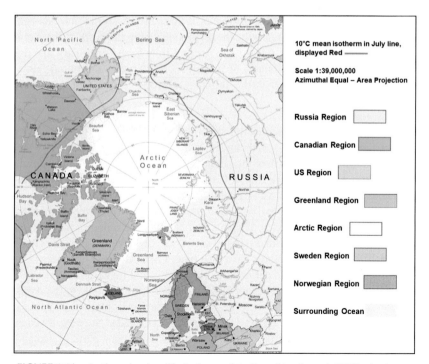

FIGURE 1.1A Arctic region (modified). *Source: http://en.wikipedia.org/wiki/File:Arctic.svg.*

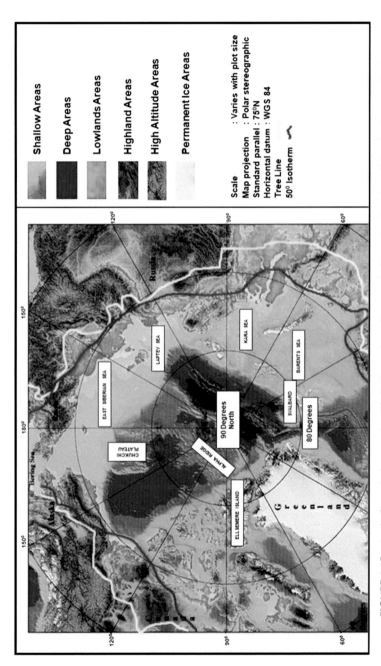

FIGURE 1.1B Arctic Ocean bathymetry (modified). *Source https://www.marinebio.net/marinescience/04benthon/arcocean.html.*

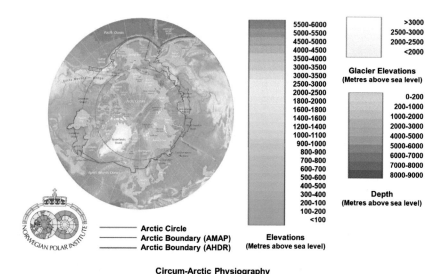

Circum-Arctic Physiography
Scale 1 : 50 000 000

FIGURE 1.1C Circum-Arctic physiography. *Source: https://arctic-council.org/site/assets/files/4330/physiography.png.*

this subarctic region has been included in the Arctic. Generally, the Arctic is a snow-covered deserted region without human inhabitants. But it is not true. During summers, Greenland and approximately 10% of the Arctic region become snow free, and vegetation like berries and flowers can be seen growing during this period. Though during winters Sun is invisible, yet in summers from March to September the sun does shine for a few hours during the day.

As believed, the Arctic is not as deserted as the Antarctic though it is one of the least populated areas of the earth. Big cities like Tromso (Norway), Barrow (Alaska), and Murmansk (Russia) are in this region. The population of the Arctic is approximately 40 lakhs with tribes like Eskimo, Chukki, and Koriak residing in rural areas. Due to lack of vegetation, their diet primarily consists of fish and meat of seals, walruses, and auks. Locals have adapted to the environment and use available resources for sustenance. They wear heat efficient clothes made of polar bear fur and seal skin. Despite the increasing use of snowmobiles and fiberglass boats, most of the population of the Arctic uses sledges, kayaks, skies, and snowshoes as a mode of transport. The Arctic is an impassable region with difficult weather conditions. Despite the availability of regular air services from the Arctic for major cities situated on the west coast of Europe and the United States, there are limited means of transport within the Arctic region. Regular air services are also available for Japan and other eastern countries from Western Europe. They fly over the Arctic region for the economy of time, cost, and distance. Many ships also pass through the northeast and northwest parts of the Arctic Ocean during summers when the ocean is snow free.

Natural resource development, sustainable economic growth, ecosystem protection, and comprehension of the impact of climate change in the Arctic necessitate furtherance of Arctic Science. Although the Arctic has gained global interest, yet it remains one of the most difficult environments in the world for scientific research in logistic perspective and is thus one of the least studied areas. A complete set of navigation charts including basic seafloor mapping remain incomplete. The region's oceans and landscapes are critically important for global migrations of fish, whales, and birds. Nevertheless, economic development and climate change will affect these populations and thus need to be structured in detail. Similarly, the effects of thawing permafrost on global methane gas emissions and of shrinking Arctic snow; sea ice, and glaciers on global sea levels; weather patterns and fisheries remain unclear to human understanding. Development of new scientific observations including long-term monitoring and mapping programs; improved computer modeling; and the development of new technologies ranging from autonomous sampling platforms to satellite observing systems is the need of the hour.

The Arctic economic significance and the complexities of the region must be evaluated in tandem for understanding the future of the Arctic. Within the Arctic and among the eight Arctic nations, noteworthy similarities as well as variations exist. Research of economic drivers and political factors across the High North is essential to judge the economic potential of the region, perceive national security interests, and develop policies for the Arctic which are acceptable [2]. As a strategic measure, some of the countries around the Arctic Ocean like Canada, the United States, and Russia have set up their defense base headquarters on their coasts. The Arctic is considered home for an astonishing wealth of mineral resources. One of the world's largest deposits of rare-earth minerals viz. iron, gold, palladium, rubies, lead, and zinc is found in the subarctic region of Canada. Large deposits of coal mines have been found on the northern parts of Alaska, Canada, and Greenland. Large deposits of Cryolite, a mineral not found easily, are found only in the west coast of Greenland. Considerable amounts of mineral oils and natural gasses are extracted from the Arctic. Scientists and geographers are trying to find out the factors responsible for the development and sustenance of life on the Arctic despite adverse weather and climate since quite some time. Scientists are trying to find out the factors that were responsible for the adaptability of human beings under such extreme and adverse climatic conditions in the Arctic. Efforts are being made by anthropologists to save and conserve the culture and traditions of the original inhabitants of the Arctic from the invasion of modern western culture.

1. Uniqueness and significance

For more than a decade, the Arctic region has become an important subject for Indian scientists and especially meteorologists. An assumption has been made that despite the vast distance from India and the Indian Ocean, the Arctic region influences the monsoon winds. It is an unbeatable truth that monsoon is the basis of our economic prosperity. A slight variance in monsoon can bring havoc to our country. To explore the interrelationship between the Arctic and tropical monsoon winds, Indian scientists are taking a keen interest in the Arctic Ocean. Many expeditions have been undertaken in this direction in the recent past.

To save the coastal areas and islands like the Maldives, it has become inevitable to stop the dissolution and melting of the ice of the Arctic Ocean, permafrost, and glaciers. Efforts need to be made to conserve and save the natural patterns, habitats, and biodiversity of the Arctic. We can do this only when we have complete knowledge of the region. We should also be aware of the impact of human activities like tourism, transportation, mining, and other strategic planning issues on the Arctic region and ocean.

Through located on the far North of Earth, the Arctic is categorized by the harsh climate, but the physical structure is like the other regions especially the southern parts of the earth.

As a matter of fact, the Polar Regions, both the Arctic (North Pole) and the Antarctic (South Pole) are cold because they are devoid of direct sunlight. The Sun is always low on the horizon, even in the middle of summer. In winter, the Sun is so far below the horizon that it does not come up at all for months at a time. So, the days are just like the nights, cold and dark. Even though the North Pole and the South Pole are "opposites," they both get the same amount of sunlight. But the Antarctic (South Pole) is much colder (Table 1.1) than the Arctic (North Pole), this is ascribed to the geographical location of both while the Arctic is an ocean surrounded by land, but the Antarctic is the land surrounded by ocean. The ocean under the Arctic ice is cold, but still warmer than the ice. So, the ocean warms the air a bit. In contrast, Antarctica is dry and high land under the ice and snow and not an ocean. The average elevation of Antarctica is about 2.3 km.

TABLE 1.1 Average temperature of the Arctic (North Pole) and the Antarctic (South Pole).

Time of year	Average (mean) temperature in °C	
	Arctic (North Pole) (°C)	Antarctic (South Pole) (°C)
Summer	0	−28.2
Winter	−40	−60

To further elaborate, contrary to the freshwater, the density of seawater increases as it nears the freezing point and thus it tends to sink. It is generally necessary that the upper 100−150 m of ocean water cools to the freezing point for sea ice to form [3]. In the winter, the relatively warm ocean water exerts a moderating influence, even when covered by ice and therefore the Arctic does not experience the extreme temperatures as experienced on the Antarctic continent.

There are many geological formations (Fig. 1.2) in the Arctic and Sub-Arctic regions, which can be termed as an extension of the extreme climate. These structures are as follows: Shield, Lower Plains, and Cordillera.

Shield: Shields are vast regions with ancient layered rocks within. Various types of layers keep accumulating over years, decades, and centuries, as time passes. The layers get thicker with time. Many ice ages have occurred and resultantly glaciers have captured them completely. After the ice ages, the traces of glaciers remain there. At present, the rocky shield region contains many lakes and the coastal areas are full of fjords.

FIGURE 1.2   Various geological structure of Arctic region (modified). *Source: https://commons. wikimedia.org/wiki/File:Major_habitat_type_CAN_USA.svg.*

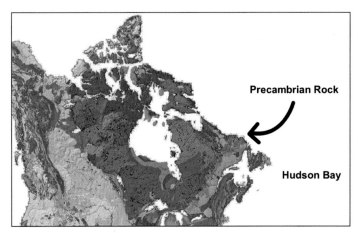

FIGURE 1.3   The Canadian Shield is a broad region of Precambrian rock (shades of red) that encircles Hudson Bay. *Source: https://en.wikipedia.org/wiki/File:Canada_geological_map.JPG.*

The Canadian Shield that lies between 60 and 100 degrees of Western latitude enclosing the Hudson Bay of Canada is far larger than the Baltic or Fennoscandian (Scandinavian) and Angara (Siberian) shields. It spans Eastern, North-eastern, and East-central Canada and the Northern portion of the upper midwestern United States. Canadian Shield (Fig. 1.3) is lowland with grass and small shrubs grow on it. Full of lakes and small rivers, this area is called as Barren Land.

Torngat Mountain is located on the eastern region of Barren land and is 2100 m high from the sea level. It accommodates various glaciers and ice caps. As there is neither barrier nor large trees on it, during winters, high winds carry its ice and snow to distant places. This affects the visibility of nearby places; reducing it to 0.8 km for 24−36 h. A very small region of the Baltic shield is part of the Arctic region. Its geographical structure is like the Southern Scandinavian mountain (Paneplained) land. The Angara Shield of middle Siberia is full of hills and valleys. The terrain is mountainous, but its map is not complete. Some of its peaks are as high as 3000 m. The north coast of Siberia is like North America but its rivers flowing toward the north are larger in quality and quantity. Three big rivers of the world namely—Lena, Ob, and Einisi are in Siberia, which flow toward the north and fall into the Arctic Ocean. North America's Mackenzie River also flows from south to north.

2. Lower plains

Lower Plains or Lowlands are in the west of Canadian shields comprising lower plains of Mackenzie and coastal areas of Alaska. The plains of Mackenzie are the expansion of Prairies on the north. Both plains are below the shield that emerged from the sea during the recent geological age.

FIGURE 1.4   Richardson Mountain ranges of Canada. *Source: https://en.wikipedia.org/wiki/Richardson_Mountains#/media/File:Dempsterhighway.jpg.*

3. Cordillera

North American Cordillera is situated in the west of the lowlands. Cordillera is part of the mountain ranges stretched from south to north of west coasts of North and South America. Cordillera comprises the Brookes Mountain range of Alaska and Richardson Mountain ranges of Canada (Fig. 1.4).

4. The Permafrost land

The Arctic region land is different from other regions of the world in two ways—firstly, due to the presence of Permafrost and secondly, because of the lack of rain. As per the scientists, permafrost is the type of soil, rock, or sediment that remained below 0°C temperatures for the past 2 years. Approximately one-fourth part of the earth comprises Permafrost. The total land of the Arctic is covered by the Permafrost layer. It has been found that the permafrost layer has spread 370 m deep in the Arctic. It is present since the ancient ice ages. During the summers, only 1 m deep layer gets melted. Due to this permafrost water is not absorbed by the earth. Either it stays on the surface or flows down to rivers. This is the main reason for a large number of lakes existing in the Tundra region. Lots of migratory birds visit this region during summers. Arctic is also known as "Cold Desert." Rainfall is less than 25 cm. There are plenty of water resources like lakes, rivers, and swamps but due to harsh extreme climate, the growth of vegetation is less. Due to the prevalent permafrost and lack of rain, soil formation and erosion takes place in different manner. In the climate of the Arctic, the chemical and biological biodegradation do take place properly. The soil formation is also slow. Therefore, vegetation does not grow properly.

The depth of permafrost is 600 m in Siberia, the deepest in the world. The depth depends on the natural geological structure of the region, its climate, and its historical background—whether it remained under the sea or glaciers in the past. It has been found that the depth of permafrost is more in those regions which were free from glaciers during the ice ages. This is the reason that the depth of permafrost is less in the sub-Arctic region. Also, it did not expand continuously in this region. The permafrost is found both on bedrocks and sediments layer of soil. Due to the presence of permafrost, there is no water drainage at all. That is why small lakes are formed during summers. Besides this, the continuous erosion does not take place there. Instead, solifluction takes place. In this process, the large rocks are eroded, and resulting materials gravel, sand, and foam keep flowing rapidly to level the surface.

The processing of permafrost is not known very well to the scientists yet. In regions where permafrost is present, the behavior of materials on the earth's surface becomes abnormal. Due to repeated frosting and defrosting, the surface substances of the earth keep changing their shape and size and get accumulated in heaps in irregular manner here and there. These regions are called Polygons. The diameter of polygons may vary from less than half meter to 100 m. Various substances are found in these heaps and therefore different types of vegetation growing there.

5. Ice Lens

Apart from the Polygons, another special feature of the Arctic region is Ice Lens or Ground Ice. Ice Lens is a vertical mass of ice having length up to 15 m, diameter up to 3 m, and depth up to 15 m. These ice lenses become obstacles in the construction of roads and houses. As when the ice lenses melt, they take away the part of the road along with them. Or sometimes a part of the road digs in due to the melting of these ice lenses. To avoid such incidences, these ice lenses are covered with a layer of nonconductive materials. The ice lenses of 6−10 m length can be seen on the cliffs situated on the banks of rivers and seas (Fig. 1.5).

Approximately 60-m-thick fossil ice is found in northern Siberia. It can be the ice of a lake or glacier buried under the earth's surface.

FIGURE 1.5 Ice Lens in Arctic (modified). *Source: https://upload.wikimedia.org/wikipedia/commons/6/61/Melting_pingo_wedge_ice.jpg.*

6. Glaciers and ice caps

It is a common belief that the whole of the Arctic region is covered with ice. But it is not the truth. Only two-fifths (2/5) of the area is covered by ice for the whole year; other parts are free from ice for most of the year. Like the other parts of the world, glaciers are formed only in those regions where the ice does not melt completely during the summers. After the snowfall, the accumulated snow slowly converts into ice. In most of the Arctic regions where the temperatures are below freezing point, the snow keeps converting into ice.

In northwest Greenland, no signs of glaciers were found even after digging 360 m below. Approximately 800-year-old ice layers were found there, which enabled the geographers to assume the glaciations of the past eight centuries. Like other parts of the world, the height of snow line varies in different parts of the Arctic. The snow line is an imaginative height at which the glaciers can exist and sustain. The biggest glacier of the northern hemisphere is in Greenland. Expanding 2510 km from north to south, it is 960 km broad and 1700 m thick. It covers 85% of the total area of Greenland. The snow line of Greenland is at 100—200 m high. In the Arctic, glaciers are not formed in coastal areas. The glaciers have slid down from the high mountains to the coasts. The geographic shapes and structures of the glaciers of Greenland resemble a group of islands connected by the thick sheet of ice with each other. The geologists assume that if the ice of these glaciers gets melted due to any reason, these islands will emerge as high as 1000-m-high from the sea level. Generally, the pieces of glaciers keep sliding toward the Atlantic sea and fall into it. These ice bodies keep sliding along the Labrador waterway toward the south till they get dissolved completely into the sea (Figs. 1.6A and B).

There are many glaciers in the islands of the northern parts of the Arctic region. Approximately 90% area is covered by glaciers in Svalbard. Likewise,

FIGURE 1.6A   A panoramic view of Konebreen Glacier, as seen from cruise ship. *Source: Shabnam Choudhary, personal communication.*

FIGURE 1.6B   Portage Glacier is a glacier on the Kenai Peninsula of the United States state of Alaska. *Source: Shabnam Choudhary, personal communication.*

Franz Joseph group of islands is also covered by glaciers. Some of the islands are totally covered by glaciers and snow caps. Considerable areas of the islands located in the North region of Russia are covered with snow caps, but some parts are totally free of snow. There are many glaciers in the Ural Mountain range. The glaciers surrounding the north Pacific are mostly concentrated in Alaska. Most of the glaciers in South Alaska are of Alpine type and are some of the beautiful glaciers of the world.

These include beautiful glaciers located in valleys, high mountains, and piedmont glaciers also. Hubbard, Seaward, and Malaspina are some of the remarkably beautiful glaciers. Glaciers of smaller size are also found in Elysian mountain ranges. Most of the glaciers in the Arctic region are shrinking except Greenland. Therefore, most of the earlier ice-covered areas have become ice-free. Records of the past 1000 year's glacier movements have been kept safe in Iceland. As per the records, during the 10−14 centuries, the movement of glaciers was considerably less but after that the movement was faster. The maximum movement was during 1650. After a brief slow down, the glaciers again started moving ahead till 1850. After 1890, the movement is backward.

7. The Arctic climate

The Arctic climate has varied significantly in the past as recently as 55 million years ago. During the Palaeocene−Eocene Thermal Maximum, the region

reached an average annual temperature of $10-20°C$ [4]. Quaternary glaciations, pushed the Arctic Ocean to a polar climate characterized by persistent
cold and relatively narrow annual temperature ranges. The temperature of the
surface of the Arctic Ocean is fairly constant, near the freezing point of
seawater. As we move toward the north or south of the equator the climate
starts getting cold. As the Arctic is located in the northmost part of the earth, it
has the coldest climate. The temperature dips to $-40°C$ during winters, in
nights in the Arctic.

The Arctic is located at the north of the Arctic Circle (66 degree 33′N). It is
the region where the average temperature for the warmest month (July) is
below $10°C$. The earth is tilted by 66 degree 33 angles on the axis in this circle.
Therefore, the region never gets the direct rays of the sun and, consequently,
the Arctic region does not receive enough heat. Here whatever heat is acquired
by the earth from the solar energy, it sends back to the space. Due to the high
Albedo of the ice, the artic area becomes colder and chilly. The temperatures
also vary at different regions of the Arctic during winters. The temperature
depends on the geological structure, the cold or temperate water bodies
flowing and the winds blowing in a region. Although the average temperature
of Queen Elizabeth Island is measured $-34.4°C$ and the temperature at Baffin
Islands is $-23.3°C$, the temperature in the mid-Arctic is measured $-40°C$
during January.

The most peculiar thing is that the coldest place of the northern Hemisphere is not located in the Arctic region, but a place named Verkhoyansk
located, in the south of the Arctic, in northeast Siberia. The average temperature of this place is $-40°C$ in the month of January. Sometimes the minimum
temperature is measured up to $-69°C$ also. There is no season called summer
in the Arctic. Generally, the average temperature remains $10°C$ and during the
end of July and beginning of August, it reaches $21°C$ in some parts. The ice of
most of the Arctic region melts during this period. The average rainfalls in the
Arctic including the snowfall is $15-25$ cm, which is considerably low. It rains
in summers and fog is formed. Despite less rain, the land of the Arctic is
mostly wet due to the low rate of evaporation. The upper layer of ice on the
permafrost melts during summers but it is neither drained nor absorbed in the
earth.

Storms are common during winters in the Arctic. The storms rise in places
with low pressure of air like the Aleutian region and Iceland. The low-pressure
region extends from East Siberia to the Bay of Alaska. The low-pressure areas
of Iceland include Central Canada, approximately half Arctic, some parts of
north Atlantic and north Europe. The storms arising from these regions move
from North West to South East direction. Based on climate, the Arctic can be
divided into two parts—the region with snow caps and Tundra. In the regions
with snow caps, the temperature never rises above $0°C$ in any month but in
Tundra region the temperature goes above $0°C$ for a month or more, though
never rises above $10°C$.

Some geographers have differently divided the Artic as a Coastal region and Continental region. In the coastal areas of the Atlantic and Pacific Oceans, the temperature goes down occasionally but the snowfall is more. On the contrary, in continental regions like North Canada and Siberia, the temperature falls considerably but the snowfall is less. Between these two types of regions, there are different small and big regions that are the exception in terms of climate. These areas include the coastal areas with extreme winters.

In the far northern regions of the Arctic, the winters start comparatively early during August but in southern parts near Tree Line, it starts from September. The temperature in the environment starts falling rapidly till December. After that during January, February, and March, it remains at around $-20°C$ throughout the season. There is minimum snowfall in the North Canadian Islands and the North Greenland region of the Arctic. The region receives less than 10 cm rain in the form of snowfall. The winters are generally stormy season for Coastal Arctic, Aleutian, Coasts of South Eastern Greenland, Island, and European Arctic. There are heavy snowfall and rain during this season. The temperature occasionally falls below $-25°C$ and never gets extremely cold. The temperatures never vary during summers in various parts of the Arctic. In Southern border areas of the Arctic, the temperature reaches $10°C$ while the climate remains temperate in internal and continental areas. The maximum temperature reaches up to $21°C$. The sun shines continuously, and the weather is "calm," but the summer season ends with rains with heavy winds.

In the coastal areas, the summers are comparatively cold till the water remains under the ice. The temperature reaches $7°C$ in southern regions, but it rises to $5°C$ in the north. The Arctic is always covered by clouds and fog during this part of the year. Such dark and thick clouds are found nowhere else in the world. The summers are known for the sudden change in weather. Sometimes the calm and pleasant climate changes into stormy weather with $0°C$ temperatures. The sky covers with cloud and the fog engulfs the surroundings. Thereafter, the temperature decreases by $10-15°C$. All this happens at the regional level. The accumulated ice on the upper level of permafrost starts melting during summers. There is no single month in the Arctic without cloud and snowfall. There is one significant fact about the climate of the polar region. During the 20th century, a remarkable change has occurred due to which the glaciers are shifting backward. The expansion of the sea around Iceland, Svalbard, and southwest Greenland has reduced. The ice deposits have become thin. The birds, animals, and especially fishes have shifted to those North regions where they never existed earlier. These changes have a considerable impact on the economy of Greenland. There are noticeable changes in the economy. The seal was the basis of the economy of Greenland. It has been replaced by fishes. The fishes of the Atlantic are found in oceans at 70 degrees North latitude.

Due to such changes, the independent economy of Greenland has become dependent on international trade. One of the significant changes that have occurred is the sheep trade. Nowadays the sheep trade has become a successful business. The change in the climate of the Arctic has manifested as an increase in the temperature. Though there is a rise in temperature in every season and region, but there is a considerable rise in the temperature of winter. This increase started from the beginning of the 20th century, and it was highest during the decade of 1930s. After that, it fell in some parts of the Arctic but started rising again after 1960.

8. The Arctic hydrological cycle

Water in the atmosphere is available in different forms of vapor, liquid, and ice. The total water content in the atmosphere is approximately 13,000 km^3 [5] of which 200 km^3 over the Arctic [6].

The atmosphere contains water in the forms of vapor, liquid, and ice. The total water content in the atmosphere is approximately 13,000 km^3 [5], of which 200 km^3 over the Arctic [6].

The hydrologic cycle is a conceptual model that describes the storage and movement of water between the biosphere, atmosphere, lithosphere, and hydrosphere (Fig. 1.7). Water on this planet can be stored in any one of the following reservoirs: atmosphere, oceans, lakes, rivers, soils, glaciers, snowfelids, and groundwater. Various physical processes of evaporation, condensation, precipitation, infiltration, deposition, runoff, sublimation, transpiration, melting and groundwater flow surface runoff, and subsurface flow are responsible for water to move from one reservoir to another.

The oceans supply most of the evaporated water found in the atmosphere. Of this evaporated water, only 91% of it is returned to the ocean basins by way of precipitation. The remaining 9% is transported to areas over landmasses where climatological factors induce the formation of precipitation. The resulting imbalance between rates of evaporation and precipitation over land and ocean is corrected by runoff and groundwater flow to the oceans. The planetary water supply is dominated by the oceans approximately 97% of all the water on the Earth is in the oceans. The other 3% is held as freshwater in glaciers and ice caps, groundwater, lakes, soil, the atmosphere, and within life.

However, in the Arctic region, atmospheric humidity, clouds, precipitation, and evapotranspiration are essential components of the climate system. Increasing net precipitation increases river discharge to the Arctic Ocean [7].

However, in the Arctic region, atmospheric humidity, clouds, precipitation, and evapotranspiration are essential components of the climate system. Increasing net precipitation increases river discharge to the Arctic Ocean.

Water vapor in the Arctic atmosphere has a residence time of about a week, compared to a decade for freshwater in the Arctic Ocean [8], and thousands of

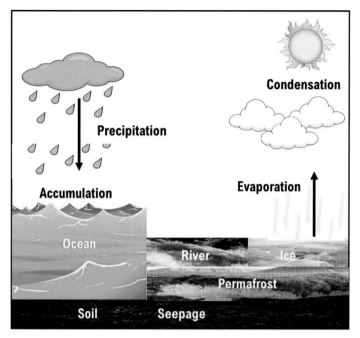

FIGURE 1.7 Schematic diagram of Hydrological cycle.

years for ice sheets and glaciers. Atmospheric moisture, clouds, and precipitation simultaneously affect and are affected by the recent rapid climate change in the Arctic (Fig. 1.8).

During recent decades, specific humidity and precipitation have generally increased in the Arctic, but changes in evapotranspiration are poorly known. Trends in clouds vary depending on the region and season.

Many efforts have been made to evaluate the freshwater budget of the Arctic Ocean [8–10], estimates on global evaporation [11], atmospheric water vapor transport, in the region −40 degree S to 40 degree N [12], and estimates reaching as far north as 70–75 degree N [13–15].

Although the Pan-Arctic moisture budget estimates were made for the first time [16] and subsequently, the synthesis of the large-scale freshwater cycle of the Arctic, including its atmospheric, oceanic, and terrestrial components during 1979–2001, was also attempted [6].

9. Sources of freshwater

Most of the Arctic region is covered by different sources of freshwater. Even in regions with less rainfall, there are plenty of sources of freshwater. There are mainly three types of sources:

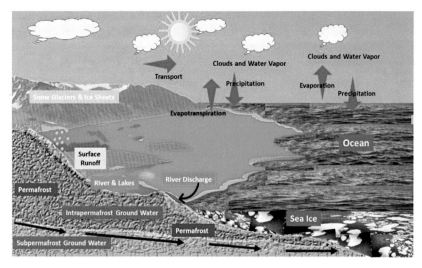

FIGURE 1.8   A view of the Arctic hydrological cycle showing important links among land, ocean, and atmosphere.

  i. Running water (rivers and canals)
 ii. Permanent water sources (lakes, ponds, etc.)
iii. Marsh

Running sources include large rivers that flow toward the north and fall into the Arctic Ocean. These rivers are Severna, Pechora, Ob, Einisi, Lena, Kolpama of Siberia, and Mackenzie and Yukon of North America. Ob, Einisi, and Lena are the big rivers of the world. Besides water, these rivers carry heat, nutrients, sediments, and pollutants with them and affect the environment of the Arctic region. These rivers have heavy current.

In addition, there are slow-flowing small rivers in the Arctic region. Most of them are found in the Tundra region. These rivers are seasonal in nature. They receive water during summers due to the melting of ice. They get frozen during winters. Therefore, they contain comparatively less nutrients. Hence, a smaller number of zooplanktons and phytoplanktons are developed in them. The sources of freshwater range from big lakes to small ponds. These get frozen during winters. By the time the ice starts melting, the summers are over. There is not enough time for water to get warm. Therefore, these water bodies are less productive as a suitable temperature is not reached for aquatic life to sustain. Marsh is ever prevailing in the Arctic. It will not be an exaggeration to say that marsh exists as the basic character of the Arctic. Approximately 11% of the total Arctic that is, 35 lac sq. km. the region comprises marshland. These are formed due to the water retention on the earth. The "*bog*" type of marshland is less productive with a more carbon deposit level, but the "*fen*" marshlands are more productive. The biodiversity of marshland is richer than the freshwater bodies.

Different types of moss are found in them. Besides these, migratory birds visit these areas and lakes of the Arctic during summer. There is a considerable impact of heat, rain, permafrost on the freshwater bodies of the Arctic.

10. The barometer of global climate

The Arctic is gaining importance during the past few decades and has become a center of attraction for scientists from all parts of the world including India. It is the *"barometer of the climate of the whole world"* by meteorologists as the impact of environmental changes all over the world is most clear and visible in this region. The effects of global warming can be felt and observed clearly on the ice of land and water of the Arctic. There is a rise in temperature from 2 to 3°C. The winter temperature is increasing. The ice has started melting before the stipulated time. The icebergs are melting. The impact of this warming will not be restricted to the Arctic region only. The melting of glaciers of Arctic and Greenland will not only raise the water level of the Arctic Ocean but also shall be resulting in the raised water level of other oceans of the earth as well. The areas below sea level may probably submerge into the water. The rainfall will increase, and snowfall will decrease. According to the report of the Intergovernmental Panel on Climate Change based on the B-2 Emission Scenario, the average temperature of the Arctic will increase up to 1°C by 2020, 2−3°C by 2050, and 4−5°C till 2080.

With the increase in the environmental temperature, the permafrost of the Arctic will start dissolving. The dissolution of permafrost will result in the emission of methane and that will add to again increase in environmental temperature. The melting permafrost might result in water logging and consequently convert various areas into marshlands. It will adversely affect the flora and fauna of the region. The ozone layer of the stratosphere which is the safety system of the nature against the ultraviolet rays has gradually developed a hole. This dreaded consequence of human actions came to light during the 1980s. The Ozone hole was detected in the sky of the Arctic continent. It has been observed that even in the stratosphere of the Arctic region, the ozone has depleted up to 45% of the original. The movement of thermohaline winds emanating from the southern region of the Arctic Ocean and the Northern region of the Atlantic Ocean is the major factor that controls the weather worldwide. Even a minor diversion or change in the movement can affect the weather and biodiversity of not only the Arctic Region but also of the whole world. To study the impact of man-made errors on the environment in the future, the Arctic is the most suitable place. Scientific organizations from various countries have established meteorological centers in the Arctic region for this purpose. Due to the global pattern of wind movement and other factors, the Arctic stratosphere is more humid than the Antarctic stratosphere. The ozone depletion fluctuates due to this reason.

The impacts of global warming in the Arctic and on the other parts of the world are described in the chapters to follow.

References

[1] https://nsidc.org/cryosphere/arctic-meteorology/arctic.html.

[2] https://nosi.org/2018/07/27/what-will-the-future-hold-for-arctic-economies/.

[3] NSIDC sea ice. Archived from the Original on 17 January 2010. Retrieved 10 February 2010.

[4] C.J. Shellito, L.C. Sloan, M. Huber, Climate model sensitivity to atmospheric CO_2 levels in the Early-Middle Paleogene, Palaeogeogr. Palaeoclimatol. Palaeoecol. 193 (1) (2003) 113—123.

[5] P.H. Gleick, Water resources, in: S.H. Schneider (Ed.), Encyclopedia of Climate and Weather, Oxford Univ. Press, New York, 1996, pp. 817—823.

[6] M.C. Serreze, A.P. Barrett, A.G. Slater, R.A. Woodgate, K. Aagaard, R.B. Lammers, M. Steele, R. Moritz, M. Meredith, C.M. Lee, The large-scale freshwater cycle of the Arctic, J. Geophys. Res. 111 (C11) (2006).

[7] T. Vihma, J. Screen, M. Tjernström, B. Newton, X. Zhang, V. Popova, C. Deser, M. Holland, T. Prowse, The atmospheric role in the Arctic water cycle: a review on processes, past and future changes, and their impacts, J. Geophys. Res. Biogeosci. 121 (3) (2016) 586—620.

[8] E.C. Carmack, M. Yamamoto-Kawai, T.W. Haine, S. Bacon, B.A. Bluhm, C. Lique, H. Melling, I.V. Polyakov, F. Straneo, M.L. Timmermans, W.J. Williams, Freshwater and its role in the Arctic Marine System: sources, disposition, storage, export, and physical and biogeochemical consequences in the Arctic and global oceans, J. Geophys. Res. Biogeosci. 121 (3) (2016) 675—717.

[9] H. Mosby, Water, salt and heat balance of the North Polar Sea and of the Norwegian Sea, Geophys. Nor. 24 (11) (1962) 289—313.

[10] K. Aagaard, P. Greisman, Toward new mass and heat budgets for the Arctic Ocean, J. Geophys. Res. 80 (1975) 3821—3827.

[11] M.I. Budyko, Atlas of the Heat Balance of the Globe [in Russian], plates 27—52, Hydro-meteorological Service, Moscow, 1963.

[12] E. Palmen, L.A. Vuorela, On the mean meridional circulations in the Northern Hemisphere during the winter season, Q. J. R. Meteorol. Soc. 89 (1963) 131—138.

[13] V.P. Starr, J.P. Peixoto, A.R. Chrsti, Hemispheric water balance for the IGY, Tellus 17 (1965) 463—472.

[14] L.R. Rakipova, Heat transfer and general circulation of the atmosphere, Izv. Atmos. Oceanic Phys. 2 (1966) 983—986.

[15] A.H. Oort, The observed annual cycle in the meridional transport of atmospheric energy, J. Atmos. Sci. 28 (1971) 325—339.

[16] A.H. Oort, Year-to-year variations in the energy balance of the Arctic atmosphere, J. Geophys. Res. 79 (1975) 1253—1260.

Chapter 2

Polar lights and midnight sun

Various unusual natural happenings, which are not found anywhere except Antarctic Continent, have been observed in the Arctic region. These happenings are miracles for the rest of the world and beyond belief. These are the Polar Lights (Arora) and Midnight Sun. Polar lights exist during the days in the summers but are visible during the nights only. Similarly, the Midnight sun shines during the night only. On North and South Polar regions, various shapes of shimmering color lights are seen across the night sky. These are not experienced regularly. The colors of the shapes are visible when they are bright generally green and yellow. This ethereal display is known as the aurora borealis (Fig. 2.1) in Arctic and sub-Arctic region and "Aurora Australis" in Antarctic. The shape of the polar light often appears to be an arc expanded vertically for kilometers and becomes dim upwards. Generally, this arc is one mile (1.6 km) from North to South but it can expand thousands of kilometers from East to West. Its average height from the earth's surface is 110 km. The color of the highest light is red and the lowest is orange in color. The high light

FIGURE 2.1 Aurora borealis over Arctic region. *Source: Photo Courtesy Shabnam Choudhary; Marika Marnela, personal communication.*

The Arctic. https://doi.org/10.1016/B978-0-12-823735-9.00018-7

looks like violet during morning and evening hours. As observed from the satellite photographs from Alaska, Canada, and especially Norway; 95% lights emanate 90−130 km above the surface of the earth and the height differs as per the shape. Till now, the height of the highest light has been found to be 1130 km and the lowest was 60 km high.

1. The shapes

The polar lights can be shaped like Drapery, Bright Spots, Arc, Arc strip, Flame or Shining Surface, or anything else but they are formed of the long thin rays spread upward for kilometers and are located parallel to the magnetic field of the earth. They glitter at intervals. The lights appear and disappear in seconds. Sometimes red particles appear in these flickering lights that disappear in minutes.

The most spectacular view of the aurora is flickering, flame-like patches and serpentine shapes writhing lifelike across the sky. All "t" rays gather round the native magnetic celestial point and flickers, flame, and rework into a shape of eagle wings and snakes writhing across the celebrities [1]. It keeps on repeating after every few seconds. Generally, the northern lights are confined to the magnetic fields near poles. The magnetic fields are expanded between 20° and 25° latitudes of the region of polar magnetic field. These fields are between 20 and 25 degrees latitudes. The Northern geomagnetic pole is in North West of Greenland, 1450 km away from geographic North Pole. Sometimes the bright polar lights can be also be seen at central latitudes. The change in magnetic field of earth also affects the Northern lights. When charged particles from the sun strike air molecules in Earth's magnetic field, they cause those molecules' atoms to become excited. The molecules give off light as they calm down.

2. Isochasm

A line connecting points on the earth's surface at which the aurora is observed with equal frequency is known as Isochasm. It is also called "isaurore" [2]. There are many assumptions about the polar lights since ancient times regarding the northern lights. The frequency of the northern lights keeps changing. Therefore, the scientists have come out with the concept of Isochasm, an imaginary line on a map or chart connecting points where auroras are observed with equal frequency as stated earlier. Isochasm is generally circular lines, and their center is geocentric pole.

3. The intensity

The intensity of these lights depends on the existence and frequency of appearance, and it is an international formula to calculate the intensity. The intensity of faint aurora like Milky Way is assumed as "I." The intensity of Northern light like the silver lining of a cloud in moonlight is assumed as "II," the intensity of Northern light like the shining of a white cotton colored cloud in full moon light is assumed as "III," and the intensity of the polar light equal to moonlight is considered as "IV." In this way, the intensity of polar lights with III intensity has been measured to be some per square microcandle. The colors of these Northern lights (auroras) with I, II, and III intensity are quite faint, and human eyes cannot differentiate these colors. The colors of auroras with IV and III intensity are yellow-green and sometimes they look like violet and red. The studies conducted by radars and photographs about these activities of northern lights differ at daily and seasonal level. The Aurora activity seems to be at its peak during late evening between 10 p.m. and 1 a.m. but can be seen in the afternoons during the polar winter above the Arctic Circle.

During equinoctial months (March—April and September—October), they are most active due to the low light pollution and crystal—clear air. The auroras occur all around the year, but we need nightfall to see it as the sun's light during the day overpowers the northern lights. The activity cycle of the auroras is quite regular. Their activities are at climax after 27 days. Similarly, after every 11 years, the activities reach their peak. This shows that there is a correlation between the Polar lights and geomagnetic movements. These Northern lights/auroras generally occur during the geomagnetic movements only. During strong geomagnetic storms these, auroras are visible at lower latitudes also. Generally, these auroras vibrate with the vibrations of geomagnetic fields. On certain occasions, these auroras produce low whistle and crackling sound and even radio waves of 3000 megacycle.

4. Reasons

There is a correlation between the Northern Lights and Sun Spots. When the Sun Spots are active, the frequency of the Northern Lights increases. Also, there is a direct correlation between geomagnetic disturbances and activities of sunspots (Fig. 2.2). When the sunspot's activities are more, the geomagnetic disturbances are most active.

We can conclude that the factors and elements responsible for these Northern lights generate from Sun. The light-weights aurora borealis aurora or northern lights are discharged particles from the sun that have the magnetic protect of earth and build light after they combine with atoms and molecules like gas and atomic number 8 gases on stepping into the earth's atmosphere [3]. These particles travel 149 million km or 93 million miles through space toward planet earth being drawn toward the earth's magnetic north and south polar regions. In other words, high activity cycles called Solar Flares the increased particles travel through space and bombard our magnetic shield to deform it slightly in shape. Some of these particles penetrate our magnetic field and meet different gases that produce the aurora northern lights that we

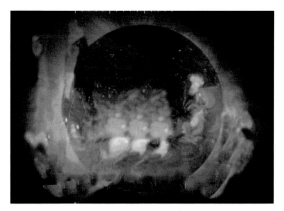

FIGURE 2.2 Sunspot activity (modified). *Source: https://en.wikipedia.org/wiki/Solar_cycle#/media/File:Solar_Cycle_Prediction.gif.*

see. However, till date, the scientists are not able to find out that how these charged particles reach the higher layers of atmosphere and why they do not get reflected toward the space. As per the opinion of the scientists, the instability of protons and electrons present in atmospheric gases could be the reason behind this.

To find out the structure of the Northern Lights (Fig. 2.3), the Scientists have tried to analyze their spectrum. It has been found that when the solar wind particles enter the earth's atmosphere they collide with molecules of nitrogen and oxygen that renounce their agitated energy in the form of light. Oxygen typically produces green and yellow light while nitrogen produces reds, violets, and occasionally blue. Violets typically form a border around curtains of green aurora shapes in lower altitudes. The occasional red color is the result of atomic reactions in lower altitudes.

5. The midnight sun

The unique phenomena of perihelion take place in the Arctic region. It is a physical phenomenon that consists of a bright spot to at least one or each side of the Sun. Two sun dogs usually flank the Sun inside a 22 degrees halo. Perihelion also known as the sun dog/mock sun/phantom sun is a member of the family of halos, caused by the refraction of sunlight by ice crystals in the atmosphere. Sun dogs typically appear as a pair of subtly color patches of light, around 22 degrees to the left and right of the Sun, and at the same altitude above the horizon as the Sun. They can be seen any place within the world throughout any season, however they do not seem to be continually obvious or bright. Sun dogs are unremarkably caused by the refraction and scattering of sunshine from plate-shaped polygonal shape ice crystals either suspended in high and cold cirrus or cirrostratus cloud clouds or drifting in

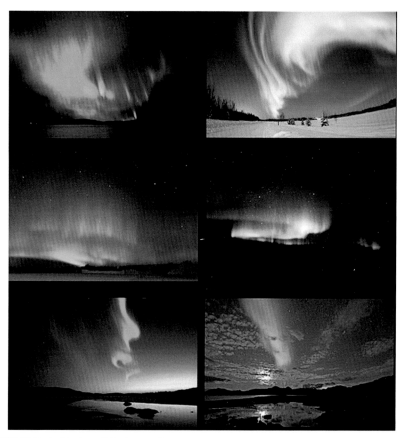

FIGURE 2.3 Northern light. *Source: https://commons.wikimedia.org/wiki/File:Aurora_Borealis_and_Australis_Poster.jpg.*

chilling wet air at low levels as frost snow (Fig. 2.4). The crystals act as prisms, bending the sunshine rays passing through them with a minimum deflection of 22 degrees. Because the crystals gently float down with their giant polygonal shape faces virtually horizontal, daylight is refracted horizontally, and sun dogs are seen to the left and right of the Sun. Larger plates wobble a lot of, and therefore turn out taller sundogs. Sun dogs are red-color at the aspect nearest the Sun; farther out the colors grade through oranges to blue [4].

The colors overlap considerably and are muted, never pure or saturated. The colors of the sun dog finally merge into the white of the solar halo (if the latter is visible). Constant plate-shaped ice crystals that cause sun dogs are also responsible for the colorful circum zenith arc, which means that these two varieties of halo tend to cooccur.

FIGURE 2.4 A sun dog (or sundog), mock sun or phantom sun, scientific name Parhelion (plural Parhelia), is an atmospheric phenomenon that crates bright spots of light in the sky, often on a luminous ring or halo on either side of the sun. *Source: https://commons.wikimedia.org/wiki/File: Fargo_Sundogs_2_18_09.jpg.*

The latter is often missed by viewers, as it is located directly overhead. Another halo selection typically seen in conjunction with sun dogs is that the 22 degrees halo that forms a hoop at roughly identical angle from the sun because of the sun dogs, therefore showing to interconnect them. As the sun rises higher, the rays passing through the plate crystals are increasingly skewed from the horizontal plane, causing their angle of deviation to increase and the sun dogs to move farther from the 22 degrees halo, while staying at the same elevation.

Generally, it is dark everywhere during midnight except full moon, but there are certain regions on the earth where the sun shines on the sky despite night. The condition is that the sky should be free of clouds. This is a common phenomenon in the north of earth till North Pole and in the south of earth till Antarctic Circle. In the Antarctic continent between Antarctic Circle and South Pole, there is no human settlement, but in the north between Arctic Circle and North Pole there are many cities and human settlements. On 21st of June every year, the sun at the Arctic Circle shines throughout the night. This occurrence is called the "Midnight Sun." The more we move toward north, the day duration increases and at North Pole it reaches to 186 days. In Canada, Alaska (United States of America), Greenland, Norway, Finland, and North parts of Russia, the Sun keeps shining day and night for 73 days (1/5 year) as these regions are located North of Arctic Circle. In the Svalbard group of islands, which is the most populated region of Europe, the Sun does not set from 19th April to 23rd August every year. And, on the contrary from 20th October to 18th April, the Sun does not come out of horizon and does not shine in the sky.

The more we go toward the North of the earth during summers the duration of the day and Sun increases and in North Pole the sun keeps shining continuously for 6 months. There is night for 6 months during winters. Thus, there is one day and one night on the North Pole and the same is true with South Pole. As per the geographers, an imaginary line called equator goes through the mid of earth that divides it into two parts. Though this line is not visible, it affects the geographic, celestial, magnetic, and other happenings considerably. The days and lights are equal on this line but when we go toward North or South of this line, the duration of days (light) during summers and duration of nights (darkness) during winters keeps on increasing. And in the North and South poles, these durations are of 6 months. The duration of the light is called the "Polar day," and duration of darkness is known as "Polar Night."

The reason for this is the tilting of earth on its own axis. The earth is tilted at 23−27 degrees angle. Due to this reason on the Arctic and Antarctic circles during winters, the sun does not rise above the horizon for the whole day. There is total darkness for 24 h. But on the contrary during summers, the Sun does not set on the horizon and keeps shining for 24 h. Due to refraction, this Midnight Sun can be seen on the places that are located below 1 degree from Arctic Circle. Because of this reason, the Midnight Sun can be seen in regions near the North altitudes like Island and Scotland. Even in regions located at 60 and 66 degrees North latitude, the nights are not dark on 21st June and one can read a newspaper in that light. The reason being that the Sun remains below 6 degrees of the horizons at that period.

The day of Midnight Sun is celebrated as an auspicious and festive day by the natives of Arctic region. In St. Petersburg of Russia, the period from 11th June to 9th July is celebrated as White Nights. Cultural events are organized during the last 10 days of June every year. In the north parts of Alaska also, the Midnight Sun is the occasion of joy and festivities. On that day, the people sing and dance for the whole night. The markets are kept open and people take photographs of the Sun. Generally, if the sun is shining during the night, it is difficult to sleep. The tourists in Arctic generally experience such discomfort. Due to prolong darkness during Polar Nights, due to lack of natural light, many people suffer from depression. This problem can be resolved by provision of bright lights.

5.1 Significant Arctic phenomena

The Arctic is known for unique phenomena owing to special atmospheric conditions. Microscopic ice crystals that are suspended in the air are responsible for the unique travel of light and sound over distances in polar regions. Some of these phenomena are briefly outlined in the following.

5.2 Optical phenomenon

Layers of hot and cold air refract, or bend, light rays. Light bounces off the surfaces of clouds, water, and ice to create optical illusions.

5.3 Acoustical phenomenon

Hearing of noises from much further away in the Arctic is yet another unique acoustic phenomenon takes place because cold atmospheric conditions bend sound waves differently than the air at lower latitudes. The range at which sound can be heard depends on the temperature of the air, the speed and direction of the wind, and the rate at which sound energy is absorbed by the earth's surface.

5.4 Coronas

A corona appears as a ring of light that surrounds the sun or the moon, sometimes forming a luminous disk, or even a series of rings with the sun or moon at its center when light is diffracted by water vapor.

5.5 Anticoronas

On the contrary, the anticorona or glory consists of one or more colored rings that appear around the shadow cast by an observer on a cloud or in fog.

5.6 Water sky

Water sky refers to the dark appearance of the underside of a cloud layer when it is over a surface of open water.

5.7 Ice blink

Ice blink refers to a white glare seen on the underside of low clouds. It indicates the presence of light-reflecting ice that may be too far away to see.

5.8 Mirages

Arctic Mirages and other optical illusions occur due to special atmospheric conditions that bend light. A superior mirage occurs when an image of an object appears above the actual object owing to the weather condition known as a temperature inversion, where cold air lies close to the ground with warmer air above it.

5.9 Optical haze

Optical haze looks like a fog or mist, blurring objects seen at a distance. It occurs in a layer of air next to the ground where warmer air flows up and colder air descends, creating wind patterns known as convective currents. The difference in how the warm and cold air refract light causes objects seen though the layer to blur. Optical haze occurs quite frequently in the Arctic, often making it difficult to identify details in the landscape.

5.10 Halos

A halo occurs around the sun when light is refracted as it passes through ice crystals and produces a very well-defined ring of light around the sun. This is different from a corona whose ring is created when light is diffracted through small droplets of liquid water in clouds. When a thin uniform cirrostratus cloud containing ice crystals covers the sky, the halo may be in the form of a complete circle.

5.11 Fog bow

A fog bow is caused by a process like that causing rainbows, but because of the very small size of the water droplets, the fog bow has no colors.

5.12 Whiteout

Whiteout occurs when the sky and snow assume a uniform whiteness, making the horizon indistinguishable and eliminating the contrast between visible objects both near and far.

References

[1] https://www.universetoday.com/tag/aurora/page/5/.
[2] http://glossary.ametsoc.org/wiki/Isochasm.
[3] https://www.slideserve.com/herne/my-7-wonders.
[4] https://ipfs.io/ipfs/QmXoypizjW3WknFiJnKLwHCnL72vedxjQkDDP1mXWo6uco/wiki/Sundog.html.

Chapter 3

Geospace and space weather from the poles

Soon after the initiation of the "space age" geospace assumed a fundamental role as a technological tool for all countries across the world. Communications, weather prediction, navigation, and remote sensing of natural resources, etc. are based on satellite support globally. It is quite possible that the satellite systems will sustain human colonies in space. The medium in which Earth-orbiting systems operate is hostile. It is made up of high-temperature gas and corpuscular radiation of varying densities and intensities; these solar-activity controlled variations can reach proportions dangerous to orbital stability, to electronic systems performance, to shuttle and space plane, and to the life of humans in orbit. It therefore necessitates the need to predict "weather and climate" in geospace, the inhospitable regions on Earth into which industrial activity of the Arctic region has moved during the last decades.

1. Geospace

Geospace is the region of outer space near Earth, including the upper atmosphere and magnetosphere. The outer boundary of geospace is the magnetopause, which forms an interface between the Earth's magnetosphere and the solar wind. The inner boundary is the ionosphere. In other words, Geospace can be defined as the region of outer space near Earth. This includes the upper atmosphere, ionosphere as well as magnetosphere. It can be called the domain of the Sun—Earth interactions. Geospace is also known as the solar-terrestrial environment and can be defined as that region of space that goes from the solar photosphere to the mesosphere of the Earth. As the name suggests, geospace is a combination of two terms—"geo" (which when used as a prefix, denotes earth, ground, or land) and "space" (which refers to the outer space in this case, that is, the void that exists between various celestial bodies, including Earth). Low-density particles that are electrically charged, magnetic fields, and radiation environment from the Sun to the Earth's atmosphere together constitute the geospace. Storm-like disturbances that are powered by the solar wind is formed from the plasma. This may drive electric currents into the Earth's atmosphere. When this happens, the radiation belts and the ionosphere of the geospace are disturbed. Geospace is believed to be the last of the

The Arctic. https://doi.org/10.1016/B978-0-12-823735-9.00012-6

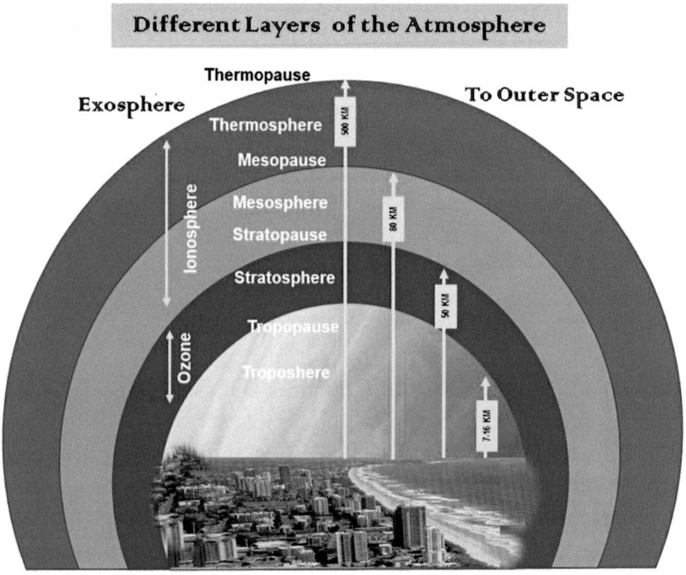

FIGURE 3.1A Different layers of the atmosphere along with the thickness of each layer.

four-physical geosphere, with the first three being solid earth, ocean, and atmosphere (Figs. 3.1A and B). As stated earlier, Geospace can be understood as having two boundaries and the Geospace field is interconnected with heliophysics, which is the study of the impact of the Sun on the solar system because the behavior and properties of the space near the earth are affected by the Sun's behavior and the weather of the space.

The uniqueness of Polar Regions for conducting geospace research has been acknowledged for decades. This is because instrumentation located at high-latitudes allows access to a natural laboratory for studying the Earth's atmosphere, its space environment, and solar-generated interplanetary structures.

The variable space-weather conditions of geospace are affected by the behavior of the Sun and the solar wind (Figs. 3.2A and B). The day-side magnetopause is compressed by solar-wind pressure—the subsolar distance from the center of the Earth is typically 10 Earth radii. On the night side, the solar wind stretches the magnetosphere to form a magnetotail that sometimes extends out to more than 100–200 Earth radii. For roughly 4 days of each month, the lunar surface is shielded from the solar wind as the Moon passes through the magnetotail. The solar wind drags out the night-side magnetosphere to possibly 1000 times Earth's radius; its exact length is not known. This extension of the magnetosphere is known as the Magnetotail (Figs. 3.2A and B). The outer boundary of Earth's confined geomagnetic field is called the Magnetopause.

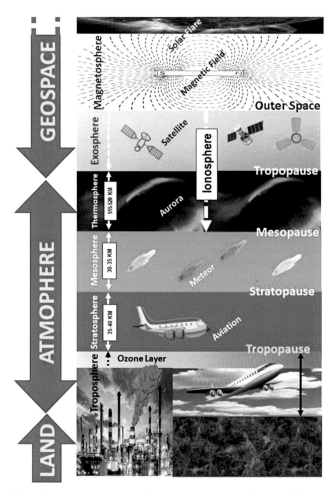

FIGURE 3.1B Systematic depiction of various activities confine to different layers of atmosphere.

Magnetosphere: It is the region around a planet dominated by the planet's magnetic field. Among all other planets in our solar system that have magnetospheres, Earth has the strongest magnetosphere.

The magnetosphere shields mother earth from solar and cosmic particle radiation, as well as erosion of the atmosphere by the solar wind—the constant flow of charged particles streaming off the sun. Earth's magnetosphere is part of a dynamic, interconnected system that responds to solar, planetary, and interstellar conditions. It is generated by the convective motion of charged, molten iron, far below the surface in Earth's outer core.

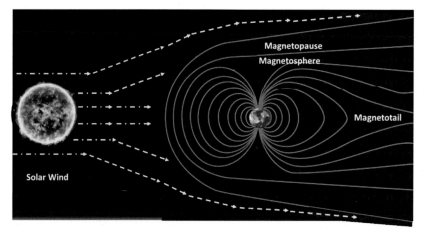

FIGURE 3.2A Pictorial depiction of magnetosphere (modified). *Source: https://www. insightsonindia.com/2020/05/04/magnetosphere/*

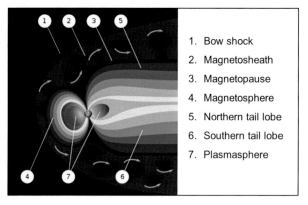

FIGURE 3.2B Schematic diagram of Earth's magnetic field. *Source: https://en.wikipedia.org/ wiki/Earth%27s_magnetic_field#/media/File:Magnetosphere_Levels.svg*

The study of the magnetosphere is essential to better understand its role in our space environment. It will unravel the fundamental physics of space, which is dominated by complex electromagnetic interactions unlike what we experience day-to-day on Earth. By studying this space environment close to home, we can better understand the nature of space throughout the universe. Additionally, space weather within the magnetosphere—where many of our spacecraft reside—can sometimes have adverse effects on space technology as well as communications systems. A better understanding of the science of the magnetosphere helps improve our space weather models.

Geospace is populated by electrically charged particles at very low densities, the motions of which are controlled by the Earth's magnetic field. These plasmas form a medium from which storm-like disturbances powered by the solar wind can drive electrical currents into the Earth's upper atmosphere. Geomagnetic storms can disturb two regions of geospace, the radiation belts, and the ionosphere. These storms increase fluxes of energetic electrons that can permanently damage satellite electronics, interfering with shortwave radio communication and GPS location and timing Magnetic storms can also be a hazard to astronauts, even in low Earth orbit. They also create aurorae seen at high latitudes in an oval surrounding the geomagnetic poles.

Lidar studies of coupling in the Arctic Atmosphere and Geospace focus on the meteorology of the middle and upper atmosphere, specifically on the contribution of small-scale processes such as waves and turbulence to the structure and composition of the atmosphere in a bid to assess the impact of meteorological processes on space weather.

Recent research in the field of Geospace is well supported by scientific instrumentation including incoherent scatter radar, satellite, optical, and radio wave instruments. Attempts are made to conduct basic physics research in the ionosphere/magnetosphere, middle atmosphere/lower thermosphere, and stratosphere/troposphere.

The research is still in the process to improve the understanding of surface-atmosphere interactions in high-latitude environments through various spatial and temporal scales using satellite, LIDAR, eddy-correlation, and other micrometeorological measurements. Such research assesses atmospheric boundary models for applications in the Arctic weather forecasts besides providing valuable hydro and small-scale meteorological data for ecosystem studies, hydrology, agricultural and forest research, and satellite remote sensing.

Observational datasets in the polar middle atmosphere are extremely valuable for understanding the polar dynamics and coupling between the lower and middle atmosphere. To improve our understanding of the mechanisms that couple solar and interplanetary processes to the terrestrial environment, such research may include, but not limited to, the study of aurora, induced electrical currents, geomagnetic fields, ionosphere electrodynamics, ion-neutral coupling, temperature and winds in the neutral atmosphere, and atmospheric waves.

The polar satellite was launched into orbit in February 1996, and it continued operations up to April 2008. It is a NASA science spacecraft, which has been designed to study the polar magnetosphere and aurorae. It was aimed to address the multifacets of Global Geospace Science.

Polar satellite (Fig. 3.3), collected images of auroras in multiple wavelengths, measured the amount of plasma used in the Polar Regions of the magnetosphere, the flow of the latter in the ionosphere and the entry of other charged particles in it and in the upper atmosphere. The entire sequence of

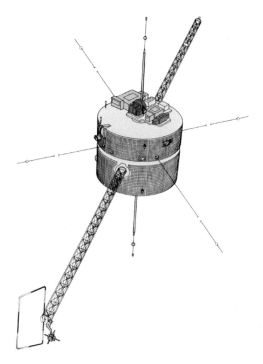

FIGURE **3.3** Polar satellite. *Source: https://commons.wikimedia.org/wiki/File:Polar_line_ drawing.jpg*

events initiated by magnetic substorms to the generation of the Aurora was observed in detail, collected data, and determined that solar storms deposited such an amount of energy in the ionosphere that it stretched to fill the magnetosphere completely.

The Polar Geospace invites the scientific communities who investigate the Antarctic and Arctic geospace by means of ground-based/space-borne obser-vations including radio probes, theories, and modeling. Various scientists are studying the neutral and/or the ionized part of the atmosphere, from the lower to further upper regions such as the magnetosphere.

It is well established that human lives and social structures are sensitively affected by the change of geospace disturbed by solar activities over the Arctic region. Scientific investigations in the contemporary fields of Arctic Geospace suggest that the geospace in the Arctic is not only linked by magnetic field lines but also by atmospheric dynamical processes including circulation and waves, which further affect the global atmosphere. Studies in the diversified domain of Arctic Geospace include, but are not limited to, the study of Space, the Sun–Earth relations, and the impact of Space Weather on critical opera-tions. The observations of the Earth from the space exploit the radio spectrums including SAR imaging, satellite altimetry, and weather satellites. Arctic atmospheric space weather system interaction is important for further developing the space weather concept and space technologies, as well as for investigating the problem of global climate changes. The observatories near Tromso and at Svalbard (Norway) are equipped with several unique facilities for environmental monitoring and diagnostics of the atmosphere— ionosphere—magnetosphere system.

Having realized the scientific importance of Arctic Geospace and space weather Norwegian and Ukrainian researchers have evolved many outreach programs to connect with the people. Such programs are focused not only for mass awareness on the development and dissemination of fundamental and applied knowledge of the properties of the near-earth space environment but are also aimed to implement the modern scientific theories and diagnostic techniques for the educational process and training of highly qualified specialists in the areas of geospace research works and development of space weather concept.

2. Space weather

Sun being the primary source of energy of planet Earth causes severe fluctuations in the geomagnetic field and disturbs the magnetosphere when the disturbances from the Sun reach the earth's atmosphere. Geomagnetic storms/ substorms cause mainly two types of effects on the Earth's ionosphere/thermosphere namely, charged particle-induced effects and electrodynamical effects are involved to change in the magnetic and electric fields of the magnetosphere.

Space weather refers to conditions in geospace that are controlled by solar activity and which can cause disruption of satellite operations, communications, navigation, and electric power distribution grids, leading to a variety of socioeconomic losses (Fig. 3.4).

Different types of space weather affect technologies on Earth in a diversified manner. Solar flares degrade or block high-frequency radio waves used for radio communication during events known as Radio Blackout Storms by producing strong X-rays. Solar Energetic Particles (energetic protons) can cause electrical failure by penetrating satellite electronics. These energetic particles also block radio communications at high latitudes during Solar Radiation Storms. Similarly, Coronal Mass Ejections can degrade power grid operations by causing Geomagnetic Storms at Earth and induce extra currents in the ground. Geomagnetic storms can modify the signal from radio navigation systems (GPS and GNSS) causing degraded accuracy. People who depend on these technologies will be severely impacted by Space weather.

Usually, space weather is insignificant in our routine life. Unless the space environment is disturbed by the variable outputs of the Sun and the technologies that we depend on can be affected. Space weather disturbances are

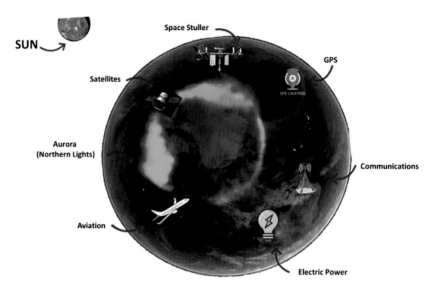

FIGURE 3.4 Schematic diagram showing the impact of the space weather causing disruption of satellite operations, communications, navigation, and electric power distribution grids.

generally caused by transient events in the solar atmosphere, which trigger disturbances in the Earth's environment. Geomagnetic storms are accompanied by a variety of ionospheres' disturbances, which are presumably caused by the intensification of the solar wind.

The ionosphere plays an active role in the complex space weather relationships that affect the distribution of plasma in the ionosphere from low to high latitude. During the periods of disturbed space weather, the ionosphere can be filled with small-scale irregularities. This problem is particularly worst at high latitudes in the region of the aurora, subaurora ovals, and the equatorial region. The aurora is a dynamic and delicate visual manifestation of solar-induced geomagnetic storms. The electromagnetic interaction of radio waves with charged particles of the ionospheres' plasma may cause the signals degradation. Space weather is of paramount significance to understand its impact on terrestrial near- and far-space environments. Recently, space weather research has got impetus having its implications both in space science and technology. A comprehensive understanding not only of the origin and evolution of space weather processes but also of their impact on technology and terrestrial upper atmosphere becomes imperative due to the presence of satellites and other technological systems from different nations in near-Earth space.

Global efforts are being made to address Sun, solar processes, and their evolution from the solar interior into the interplanetary space, and their impact on Earth's magnetosphere-ionosphere-thermosphere system. India has made significant efforts to gain a better understanding of polar space weather.

3. Ionosphere research

The Polar Regions (Arctic and Antarctic) offer a unique opportunity to serve as a natural laboratory for ionosphere research having importance for all satellite transmissions. Ionosphere Total Electron Content (TEC) data are recorded every 30 s. Polar high latitude regions are directly affected by the energy of charged solar particles and energy thus the ionosphere becomes highly fluctuation.

Severe fluctuations in the geomagnetic field and disturbance in the magnetosphere are primarily caused when the disturbances from the Sun reach the earth's atmosphere. These disturbances are usually known as geomagnetic storms/substorms that cause mainly two types of effects on the Earth's ionosphere/thermosphere—(1) Charged particle-induced effects and (2) Electro-dynamical effects are involved to change in the magnetic and electric fields of the magnetosphere.

Change in the composition of the thermosphere (95−500 km altitude) and in the composition of the upper atmosphere alters the recombination rate of the ionosphere (60 to beyond 1000-km altitude) and drastically changes the ionosphere electron density.

Various factors such as "ionization process" depend on the sun's activity and its zenith angle to affect the TEC fluctuations over the polar region. Total electron content is very variable with time and season. Several studies have demonstrated that TEC strongly depends on solar activation [1−5]. Nowadays GPS measurements are commonly used to investigate the structure and dynamics of the ionosphere.

A study of the diurnal and seasonal variation of the TEC of the ionosphere by using Geostationary satellite beacon signals had been done [6,7]. Space weather over equatorial as well as high latitude regions has been studied for a variety of purposes [8−14]. The ionospheres' parameters are mainly under the control of solar geophysical conditions (Space Weather) and it is greatly affected by geomagnetic storms especially in the high and low latitude ionosphere. The effects of space weather over high and low latitude Ionosphere are significant. At high latitudes, the contribution due to the particle fluxes becomes important. The solar wind is a significant energy source of the entire ionosphere at high latitudes. The geomagnetic field forms an obstacle to the solar wind flow, by intense interaction between the solar wind and the magnetosphere, mass, momentum, and energy is transferred from solar wind to magnetosphere.

High latitude aurora irregularities are formed from the precipitation of energetic electrons along the terrestrial magnetic field line into the high latitude ionosphere. These electrons are energized through a complex interaction between the solar wind and the earth's magnetic fields, resulting in optical and UV emissions commonly known as the auroras. This phenomenon characterized the magnetosphere substorm, where associated irregularities in electron density lead to scintillations [16,17].

The large-scale convection of the magnetic field lines due to solar wind interaction that results in disturbances and the precipitating particles that come from the collision retardation of magnetosphere energetic particles as they enter the lower atmosphere are two important energy input mechanisms at high latitudes.

References

[1] A.V. Da Rasa, H. Waldman, J. Bendito, O.K. Garriott, Response of the ionospheric electron content to fluctuations in solar activity, J. Atmos. Terr. Phys. 35 (1973) 1429–1442.

[2] H. Soicher, Traveling ionospheric disturbances (TIDs) at mid-latitude: solar cycle phase dependence, Radio Sci. 23 (1988) 283–291.

[3] P.J. Van Velthoven, Medium-Scale Irregularities in the Ionospheric Electron Content, Ph.D. thesis, Technische Universiteit Eindhoven, 1990.

[4] E. Feitcher, R. Leitinger, A 22-year cycle in the F-layer ionization of the ionosphere, Ann. Geophys. 15 (1997) 1015–1027.

[5] A. Krankowski, I. Shagimuratov, Impact of TEC fluctuations in the Antarctic ionosphere on GPS positioning oczapowski St. 1,10-957, Russ. Artif. Satell. 41 (1) (2006), https://doi.org/10.2478/V10018-007-0005-5.

[6] J.E. Titheridge, Continous records of total electron content of ionosphere, J. Atmos. Sol. Terr. Phys. 28 (1966) 1135–1150.

[7] G.O. Walker, S.D. Ting, Electron content and other related measurements for a low-latitude station obtained at sunspot maximum using geostationary satellite, J. Atmos. Terr. Phys. 34 (1972) 283.

[8] J. Aarons, M. Mendillo, R. Yantosca, GPS phase fluctuations in the equatorial region during sunspot minimum, Radio Sci. 32 (1997) 1535–1550.

[9] X. Pi, A.J. Mannucci, U.J. Lindqwister, C.M. Ho, Monitoring of global ionospheric irregularities using the Worldwide GPS Network, Geophys. Res. Lett. 24 (18) (1997) 2283–2286.

[10] S. Basu, E.J. Weber, T. Bullett, W. Keskinen, M.J. MacKenzie, E. Doherty, P. Sheehan, R. Kuenzler, H. Ning, P. Bongiolatti, Characteristics of plasma structuring in the cusp/cleft region at Svalbard, Radio Sci. 33 (6) (1998) 1885–1899.

[11] J. Aarons, B. Lin, Development of high latitude phase fluctuations during the January 10, April 10-11, and May 15, 1997 magnetic storms, J. Atmos. Sol. Terr. Phys. 61 (3–4) (1999) 309–327, 1999.

[12] M. Mendillo, B. Lin, J. Aarons, The application of GPs observations to equatorial aeronomy, Radio Sci. 35 (2000) 885–904.

[13] R.P. Basler, R.N. DeWitt, The height of ionospheric irregularities in the auroral zone, J. Geophys. Res. 67 (1962) 587–593.

[14] J.M. Lansinger, E.J. Fremeuw, The scale size of scintillation-producing irregularities in the auroral ionosphere, J. Atmos. Terr. Phys. 29 (1967) 1229–1242.

[15] H.E. Whitney, J. Aarons, C. Malik, A proposed index for measuring ionospheric scintillations, Planet. Space Sci. 17 (1969) 1069–1073.

[16] L. Kersley, C.D. Russell, D.L. Rice, Phase scintillations and irregularities in the northern polar ionosphere, Radio Sci. 30 (1995) 619–629.

[17] J. Aarons, Global morphology of ionospheric scintillations, Proc. IEEE 70 (4) (1982) 360–378.

Chapter 4

Arctic explorations: historical perspective

Harsh and extreme climate has made Artic inhabitable and very sparsely populated. Its history was created by various scientists, geographers, and geologists who came to explore the Arctic under many expeditions. Many superstitions and myths have been prevailing in various parts of the world about the Arctic region. In fact, during that period there were no maps as well as means of transport and communication for access to distant and unknown regions like the Arctic.

1. First explorer—Pythus

Pythus, a contemporary of Aristotle and Alexander, was the first person who crossed the geographic boundaries during 325 BCE. There are no records available about the expeditions of Pythus to distant north for the next generation explorers. As per the prevalent tales, he started his expeditions in 325 BCE. Different historians have different opinions about the objective of his journeys, but it is a fact that he had great enthusiasm to explore the unknown world. No one doubted his knowledge as an astronomer and mathematician. He refuted many prevailing beliefs during his time. He was the one who explored that the high tides in oceans have a direct correlation with the lunar phases.

His invincible enthusiasm to explore the world encouraged him to sail to the Antarctic through the Mediterranean Sea. After the Antarctic, he sailed toward the North and reached Brittany. From the local population, he heard about the mysterious land of aurora and the midnight sun. He proceeded farther toward the north and after 6 days of sailing he reached the edge of a frozen sea. It was a region where the sun shines from 21 to 22 h during summers. The region was located between 63 and 65 degrees north latitude. The tribes of the region named the place as "sleeping land of the Sun." The information gathered by Pythus is found to be true in the present-day context also. He has mentioned a part of the land called Thule, which is still a mystery for the scientists. According to some scientists, Thule is Shetland Islands and Orkney but some of the scientists consider it as a part of Norway and Iceland. Thule is located 1120 km North from the Arctic Circle and 1480 km from the

The Arctic. https://doi.org/10.1016/B978-0-12-823735-9.00014-X

41

North Pole. It is the base station of the American Army. It has been said that Irish saints used to visit the Arctic during the 8th and 9th century but they did not settle here. The Vikings of Scandinavian countries were the first natives of the Arctic. In the next four centuries, these adventurous natives found ways for trading. In 982, they discovered Greenland and settled two colonies there. These Vikings spread to North American coasts. They reached Spitsbergen and Novaya Zemlya. While they flourished initially, these settlements eventually foundered due to changing climatic conditions and the invasion of Ice Age. They are believed to have survived until around 1450 (Fig. 4.1).

It is said that the natives of the Arctic initially settled in Northeast Siberia. Hunting was their primary source of living. They reached Hudson Bay and Baffin Islands in search of food. Proceeding through North Arctic Islands, they even reached Greenland, but the Arctic and the Arctic Ocean remained unexplored by Vikings. Maybe they never felt the need to do so. The renaissance during the 15th century filled the Northwestern Countries with energy and zeal to enter a trade with the Eastern Countries and distant places. They started finding sea routes for trade as the land routes were captured by Turks. Their search for alternate trade routes contributed to further exploring the Arctic and the Arctic Ocean.

2. Northeast Passage

In 1553, when Richard Chancellor completed his sea journey around the North Camp and reached the place known as Archdale presently, the probability of

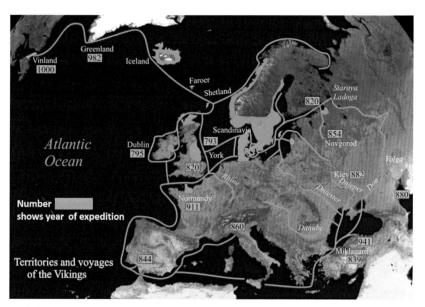

FIGURE 4.1 Viking sailors reached the White Sea to the east and Greenland and North America to the west. *Source: https://en.wikipedia.org/wiki/File:Vikings-Voyages.png.*

Northeast Passage from Europe to Eastern countries through the Arctic seemed to realize. After the journey, Richard Chancellor launched the Muscovy Company. The company promoted trade between England and Russia. The trade grew considerably and for some period Britain forgot the discovery of Northeast Passage. After 3 years, Steven Borough, the master of Chancellor's ship reached Novaya Zemlya and a Dutch trader Williem Barentz explored the total region between 1594 and 1596 BCE. Different explorations since the 15th century brought the Arctic to the notice of different rulers. By this time, the Russian traders and the fishermen had also started taking interest in various parts of the Arctic Ocean. In the 16th century, they reached the east of Taimur cape, and by the 17th century, they were able to enter the Pacific Ocean through North. At the beginning of the 18th century, the Russian explorers had strategically discovered the Siberian coasts. When the British and Dutch were struggling for the Northeast Passage, the Kazak soldiers of Russia were moving ahead toward the east through Ural Mountain. Crushing the natives of Siberia, they reached the coasts of the Pacific Ocean by the 17th century.

When the Russian Emperor Peter the Great heard about these achievements, he ordered to undertake a special expedition. Soon, the expedition was known as North Expedition (1733−42). Consequently, during the rule of the successors of Peter, the maps of Siberia's coastal river Kolyma River were prepared. And in the last, the route from Europe to the East through the Arctic was discovered by Swedish explorer Barren A.E. Nordenskiold in 1878−79 while traveling in his ship Vega. The Norwegian explorer Fridtjof Nansen observed during his journeys that the logs of wood from the forests of Siberia float through the Arctic and reach the coasts of Greenland. With an aim to win the North Pole and explore the Arctic, he started his journey on his ship. Nansen and his fellow travelers were not perturbed by the extreme climate of the Arctic, and they let their ship sail through the snow cap in the Arctic. The ship crossed three basins of the Arctic but it could not reach the North Pole as was planned by Nanson. But during his journey, he discovered and learned several facts about the Arctic Ocean that were not known to earlier explorers.

Though the trade route had been established from the Antarctic through the Arctic to Siberia's Einisei River and was being used before the First World War, it was developed after the decade of 1920s. In 1932, the ice-breaker ship Sibiryakov cleared the passage from Arkencheszk to Baring mouth and in 1934 the Ship Litke found the passage for trade from West to East. Now, with the recent development and support of ice-breaker ships, the cargo ships from Europe to East Asia cross through the Arctic Ocean directly. Their movement depends on the weather conditions. The movement of the ship is possible till the period of 4 months only. As the temperatures fall, the ocean water freezes and the ice depth is quite thick. In addition to this, the number of glaciers falling from nearby lands also increases. The accumulated ice rocks on the Arctic Ocean makes it difficult for the ships to sail through.

3. The Northwest Passage

The *Northwest Passage* is a sea route connecting the Atlantic and Pacific Oceans through the Arctic Ocean (Fig. 4.2B). Martin Frobisher made the first attempt to forge a trade route from England westward to India. In 1576−78, he took three trips to Baffin Islands and the Canadian Arctic to find the passage. The Frobisher Bay, which he discovered, is named after him. Many interesting and suspense incidents have been related to his discoveries. At that time, the maps used to have many deficiencies and errors. They used to be quite misleading and vague. Between Southern Labrador to Greenland, the presence of any geological body was not shown in these maps. Another map was prepared by Nicole Jeno in 1558 based on a sea journey during 1380. An imaginary country named Friesland was shown between Greenland and Iceland. Due to this dubious map, the discovery of the Northwest Passage could not be pursued for approximately 200 years.

4. Shipping

The Arctic is also becoming more viable as a destination and transshipping passageway. The two Arctic passages—the Northwest Passage and the Northern Sea Route (NSR) could decrease travel time between the United States, Europe, and Asia by 40% (Fig. 4.2A). This would reduce fuel consumption, save money, and speed up the delivery of goods around the world. The region has already witnessed an exponential increase in shipping traffic.

Relations report that 71 ships crossed the Arctic Ocean between Asia and Europe in 2013 compared with only five in 2010. Maritime transits through the Bering Strait increased 118% between 2008 and 2012 according to the US National Oceanic and Atmospheric Administration. By 2020, Russia predicts that Arctic shipping will increase 30-folds, and the Polar Research Institute of China estimates that Arctic shipping could carry between 5% and 15% of China's total trade value.

5. The discovery of gold mine

Martin Frobisher (Fig. 4.3) commenced his first journey in three ships namely *Gabriel*, *Michael*, and a 10-ton weighed ship *Pinnasi*. *Pinnasi* was lost in the fierce storms with four sailors. The sailors abandoned *Michael* and later returned it back. After a great struggle, Frobisher reached the South East coast of the Baffin Islands. After that, Frobisher sailed for another 100 km in *Gabriel*. He collected many things on his way; one of them being a piece of rock. The piece of rock created doubt about the availability of gold in the rocks. He forgot his aim to explore the Northwest Passage (Fig. 4.2B) and started looking for gold in the region from where he had got the piece of rock.

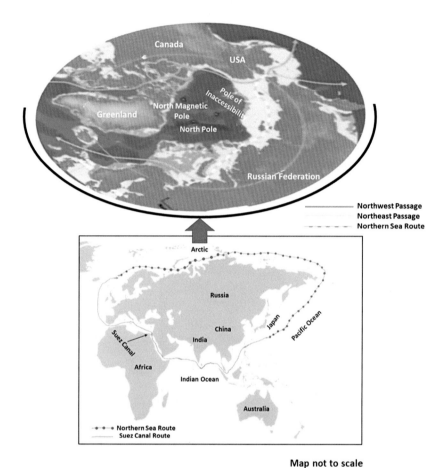

Map not to scale

FIGURE 4.2A   Map showing the Northern Sea Route and Suez Canal Route. The enlarged Arctic Region demarketing Northeast Passage, Northwest Passage, and Northern Sea Route. *Source: https://en.wikipedia.org/wiki/Northern_Sea_Route#/media/File:Map_of_the_Arctic_region_ showing_the_Northeast_Passage,_the_Northern_Sea_Route_and_Northwest_Passage,_and_ bathymetry.png.*

Frobisher repeated his journeys for another 2 years, but his only aim was to explore the gold. His last journey was quite typical. He took 15 ships and 100 sailors with him. He aimed to settle a colony near the gold mines for his men. He had also taken along readymade home structures. But, Frobisher could not find the gold mines. The ships were lost and some of them sank. He could not find the gold mines. On his return journey, he filled pieces of rocks in the ships. When he reached his native place, he could not find even an inch of gold in the pile of rocks gathered by him.

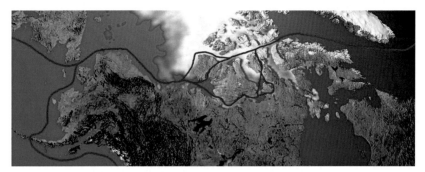

FIGURE 4.2B Northwest Passage routes. *Source: https://en.wikipedia.org/wiki/Northwest_Passage#/media/File:Northwest_passage.jpg.*

FIGURE 4.3 Martin Frobisher. *Source: https://upload.wikimedia.org/wikipedia/commons/8/8f/Sir_Martin_Frobisher_by_Cornelis_Ketel.jpeg.*

After Martin Frobisher, the other explorer of Northwest Passage was John Davis. Apart from being an expert sailor and a scientist, Davis was a humanistic person with excellent communication and behavioral qualities. John Davis reexplored Greenland from 1585 to 1587. Davis rounded the island before dividing his four ships into separate expeditions to go looking for a passage westward [1]. After reaching till Norse population, people had forgotten this region. Davis got down at the Southeast coast of Greenland and explored it till the Western coast at 72 degrees latitude.

FIGURE 4.4 W.E. Parry. *Source: https://en.wikipedia.org/wiki/William_Parry_(explorer)#/media/File:Captain_William_Edward_Parry_(1790-1855),_by_Charles_Skottowe.jpg.*

John Davis set an example of his farsightedness. To be friendly with the natives, Eskimos John Davis had taken along a group of musicians with him. He convinced his sailors to dance to the tune of the musical group. Soon the Eskimos got friendly with John Davis.

The discovery of the passage from Europe to North America through the Arctic Ocean got completed when W.E. Parry (Fig. 4.4) reached Melville Island. He was different from Robert E. Peary who had won the North Pole. He chose his passage through the West of Lancaster Sound. The polar ice became a challenge for him. In the end, Roald Amundsen (Fig. 4.5) during 1903−06 was successful to find the complete Northwest Passage. He took the round of Alaska in his small ship *Joa*, via West of Peel Sound through Southern passage of Canadian mainland. It took him 3 years to complete the whole journey. He spent two winters in William Island in experimenting with magnetism and other scientific investigations. After spending the third winter in the West Part of Mackenzie, he was able to cross through the Bering Strait in 1906. After that, Amundsen was able to win the South Pole in 1911.

A lot of adventurous and thrilling episodes occurred during the exploration of the Northwest Passage apart from the discovery of gold. The Danish explorer Janes Mack and his crew members died of scurvy at the mouth of Churchill River during winters of 1619−20. Only Mack and his two crew members could survive and return with the ship.

FIGURE 4.5 Roald Amundsen. *Source: http://en.wikipedia.org/wiki/File:Roald_Amundsen2.jpg.*

6. Endeavors by Navy

Earlier in the 19th century, the traders made attempts to explore and find the passage through the Arctic Ocean, but after defeating Napoleon the governments of European countries were left with no significant assignments. The British Army also got relaxed and started taking interest in exploring the passages. Many expeditions were undertaken to explore the passage through North West. Consequently, geographical investigations were carried out in various bays and islands of the Arctic and their correct maps were prepared. Various scientific experiments were also carried out. During such expeditions, J.C. Rous explored North Magnetic Pole at South West of the Butia cape.

John Franklin (Fig. 4.6) was the leader of the last expedition of the British Government. He proceeded to the Arctic Ocean with his two ships *Airbus* and *Terror* through Lancaster Sound in 1845 but he went missing after some time. After a continuous search by 14 ships for 12 years, it was found that he died with his crew members and the ships were destroyed. During these expeditions, various discoveries and explorations were carried out.

John Franklin (1786−1847): He went missing with his 128 crew members. On the basis of a mysterious map drawn by a four-year-old girl in her dream, Lady Franklin sent four ships to search him. In the map, the spot was marked where Franklin had died.

FIGURE 4.6 John Franklin. *Source: https://upload.wikimedia.org/wikipedia/commons/1/13/ John_Franklin_profile.jpg.*

7. Traders of Whale and Fur

It may sound strange, but the traders of Whale and Fur played a significant role in exploring various geographical regions of the Arctic. While serving the Muscovy Company, Henry Hudson had undertaken his first sea voyage through the East Coast of Greenland. The objective of his journey was to find the direct passage from Cathay to Polar Regions. After reaching 75°30′ north latitude, he turned toward the East. During his journey, he came across many Whales near Spitsbergen. He thought of hunting the whales and soon on his recommendations, many traders from many countries established enterprises of hunting the whales and obtaining oil from them. Some of the countries were England, the Netherland, Denmark, France, the United States, and others. The business of these entrepreneurs flourished for another three centuries. Because of the rapid hunting during this period in the seas around Greenland, Baffin Bay, Hudson Bay, and Buford Sea that all the resources of whales got exhausted. The trade got ended toward the end of the 20th century.

In search of whales, these traders reached unknown new regions never visited by anyone. They used to gather geographic information of the region and draw maps for their future use. Those maps were kept secret from the rivals but despite that some important facts were revealed. The coasts of Spitsbergen were explored by Dutch and British whale hunters and traders.

After that, the Norwegian whale traders also explored some more unexplored islands. The sailors also came across some islands during their search for John Franklin. The information gathered by the whale hunters proved to be useful in drafting the maps of Baffin Islands and Hudson Bay. The map of South Hampton Island was prepared by a whale hunter George Colmar and Thomas Log, and an American whale hunter had discovered Wrangle Islands.

The pair of father and son, namely William Scosarbi was famous among the whale hunters and sailors. The senior Scosarbi was the son of a farmer. Later, he proved to be an excellent sailor as well. He invented "Crows' Next" and many other useful things for sailing. He was the first person who suggested the idea of Sledge to reach the poles. He had inherited the traits of his farmer father and had also studied various subjects of science. He wrote two important books about the Arctic. The duo reached 81°12′ North of Spitsbergen in 1806. Till that time, nobody had traveled to this far of North Region. In the year 1822, they prepared maps of 69−75 degrees latitudes of eastern coasts of Greenland.

The way the whale traders contributed in exploring the coastal regions of the Arctic, the same way the Fur traders contributed in discovering the interior regions of the Arctic indirectly. The Kazaks of Siberia had laid the foundation of Fur business by hunting the animals with furs. This trade also flourished till the animals with fur became almost extinct.

8. Exploration by air

The Arctic was not only explored through road and sea routes, but balloons and aeroplanes were also used for the survey. The first attempt in this direction was made by S.A. Andry in 1897. After that, Amundsen and Lincoln Alsberth were also successful in flying over the Arctic from Svalbard to Alaska. In 1928, George Wilkinson flew from Alaska to Spitsbergen. After that, it became a regular feature. Presently, there are regular flights from Europe to distant Eastern countries over the Arctic. The significance of the Arctic has been enhanced with the use of this Great Circle route for air travel. Toward the end of the 19th century, proposals were passed to strategically establish Meteorological and Magnetic Observation Centers in the Arctic with international support during the International Polar Summits. Subsequently, the United States, Norway, Sweden, Denmark, Russia, Great Britain, Germany, and other countries set their Observation Centers in the Arctic region in 1882.

Russia was identifying Arctic and Sub-Arctic regions as "Agricultural regions of the future" but during 1937−38 Britain, Denmark, Norway, the United States, and other countries conducted various expeditions for better utilization of available resources of these regions. Under the leadership of Evan Papanin, Russia sent an explorer group to the Arctic. That explorer group spent the winter near an Ice Flow and kept floating with the flow of water for 274 days.

He covered the distance between 19 latitude and 58 longitudes during his expedition. Simultaneously, he gathered important scientific information about the water, weather, and magnetism about this vast region. During 1937–40, Russia sent its ice breaker ship *Sedov* to gather information about the above Ice Flow and discovered that there is no existence of an Island named Sunnykov. In 1938, Long Coach took photographs of Greenland from the air, and it was observed from the photographs that the popular Peary Channel is a mere Fiord Russia established its Meteorological and Radio centers in the Arctic before Second World War. In addition, it started to develop the region to avail the benefit of its natural resources. The Arctic Ocean was also used to transport arms and ammunition during the Second World War. The attempts continued even after the War. The Canadian Army took a keen interest in this region to test the vehicles that can be used on ice and Snowmobiles. They also made efforts to develop Alaska and North Canada, but the efforts were not so serious as compared to the efforts made by Russia. They also tried to get the support of the natives of the Arctic in their efforts for development.

After 1954, the United States and Russia established many research centers on Ice flows for scientific studies and observations. In 1955, a radar network was established with the cooperation of the United States and Canadian Military Organisations that could send quick signals upto 4500 sq. km from Alaska to Baffin Islands. After that, it was extended through Greenland. During the International Geostructure Year, 1957–58 more than 300 countries established their Observation Centers in the Arctic. The Arctic Institute of North America contributed majorly in the introduction of sponsored planning for the studies of the Arctic. In 1960, this Institute established a permanent Center in Devon Island for Arctic studies. This center works all throughout the year.

9. Travel through submarine

In 1958, Nautilus the atomic energy-driven submarine was able to travel below the North Pole. It was the first submarine in the world. Nautilus traveled continuously 2930 km under the water and reached the North Pole on August 3, 1958 at 11:30 p.m. But it was not the first submarine to travel below polar ice. In 1931, the British explorer Hubert Wilkins (Fig. 4.7) had tried to travel by an old designed submarine of the American Navy. He tried several times but could not reach the North Pole as the submarine was not suitable for such a challenging journey. He was the first person to fly over the Arctic from Alaska to Spitsbergen in 1928. He also tried to travel toward the North Pole by submarine in 1931 [2].

After Nautilus another atom energy-driven American submarine *Skate* traveled 3850 km below the ice and reached the North Pole in March 1960. *Skate* had set a record of traveling 4800 km continuously for 12 days below the ice. Another atom energy-driven American submarine—Sea dragon reached

FIGURE 4.7 Hubert Wilkins (1888–1958). *Source: https://upload.wikimedia.org/wikipedia/commons/f/f1/Hubert_Wilkins_1931.jpg.*

the North Pole in 1960 through Northwest Passage. It had tipped upto 100 m under a giant ice rock. Atom-driven submarines can stay under the sea for a long time and therefore can study the ice on its surface. It measures the depth of the sea more accurately and can observe the sea level, its water, ice, and its animals that are not possible for the submarines driven by steam. Nautilus and Skate had verified the presence of the Lomonosov Mountain range below the Arctic Sea. The ordinary ships had to wait for another 32 years to reach the North Pole after *Skate*. I.B. Oden and F.S. Polarstern were the first steam-operated ships to reach the North Pole. They reached the North Pole on September 7, 1991. After that, many ships traveled to the North Pole.

After the discovery of mineral oil deposits in 1960 in the North Slope of Alaska and in 1972 at Ellsmere Island of Canada, the exploration work for finding further deposits started at full pace. In the summers of 1969, the ship *S. S. Manhattan* equipped with an ice breaker and other marine scientific instruments traveled from Philadelphia to Alaska through Northwest Passage. After that, many commercial ships traveled through Northwest Passage. A huge International project "Arctic Ice Dynamics joint Experiment" was started in 1971. Its objective was to observe the impact of Icebergs on the climate of various countries of the world. Every landscape of the Arctic has been documented and photographed through satellite and not observed by sensitive equipment.

10. North Pole conquest

In the beginning, the main objective of explorers who reached the Arctic was not to conquest the North Pole. In fact, the core objective of the expeditions was to find a sea route to Eastern countries through the North Pole. Therefore, even after 200 years of the first attempt by Hudson, the British Government did not organize an expedition for the conquest of the North Pole. In the year 1773, one expedition was sent to the North Pole but it could only reach up to Spitsbergen. In 1827, W.E. Parry was the first person who used Sledge suggested by William Scoresby. W.E. Parry reached till 82°45' north latitude.

All attempts were made through the sea route between Greenland and Spitsbergen. This route was not suitable to reach the North Pole because of a cold-water stream that used to pass from north to south; it was full of ice flowing through. After many attempts by Franklin during his expeditions, a Western route to Greenland was found. The route was first adopted by Hayes in 1860 when he attempted to reach the North Pole with his ship "United States." There was a strong assumption that the North Pole is in ice-free ocean and one can reach there by breaking the packed ice only. Hayes strongly believed in this assumption. Unfortunately, he had to face extremely heavy ice during his voyage and he failed in his expedition.

In the year 1871, Charles Francis Hall (Fig. 4.8) succeeded in his expedition and he reached the North end of Lincoln Ocean Channel (82°11' North). During the expedition, he explored both coasts of the Channel and prepared its maps also. Despite this, the luck did not favor Francis Hall and he died during winters. His ship "Polaris" got stuck in the ice and later flew toward Smith Sound in the South. Till that time, it was totally damaged. The other crew

Charles Francis Hall

FIGURE 4.8 Charles Francis Hall. *Source: https://en.wikipedia.org/wiki/Charles_Francis_ Hall#/media/File:Charles_Francis_Hall_only_known_photo_(cropped).jpg.*

members who were traveling with him in "Polaris" comprised an Eskimo woman with a 2-month-old child who parted away. They spent the whole winter in an ice floe floating toward the South. At last, they reached Labrador in April 1875 and a whaler rescued them.

During 1875–76, a British expedition crew reached the Lincoln Sea in two ships "Alert" and "Discovery" under the leadership of Capt. G.S. Norge. Later "Alert" pulled on at Cape Sheridan and "Discovery" at Lady Franklin Bay in South due to winter. On arrival of spring, two groups came off the ships and proceeded on Sledges to explore the coasts of Ellesmere and Greenland. Only one of the groups reached $83°21'$ north. During this period, the Spitsbergen Passage was abandoned for some time. During 1869–70, a German Expedition group under the leadership of Carl Coldway reached the East Coast of Greenland in the ship "Germania" and explored Cape Bismark. Another ship of the expedition group "Hansa" got missing and was crushed under the ice. Fortunately, the sailors escaped and were rescued to a safe place on flowing ice. The world-famous Swedish explorer A.E. Nordenski conducted two expeditions to the North Pole through Spitsbergen Passage. The first expedition was conducted by a ship in 1868 and the second by a Readier Sledge in 1873. In 1879, another expedition group under the leadership of Lt. Commander G.W. Day Long took a new route in the ship "Zenith."

According to the prevailing assumptions "Wrangler Island is a huge part of land spread up to North." The above crew led by Lt. Commander G.W. Day Long planned to travel along the coast of Wrangle as far as possible toward the North and then reach the North Pole by Sledge. According to the plan, this expedition group passed through Bering Strait, but the ship "Zenith" got stuck in ice and kept floating toward the West for another 22 months. It crossed near the Wrangle Islands also. One achievement of the journey was that the expedition group came to know the actual expanse of the Island. After that, the "Zenith" sank into sea near New Siberian Islands. The group traveled by foot till the Delta of Leena River. The journey was very tough and cumbersome. Many group members lost their lives during the journey. After 2 years, the wreckage of "Zenith" was found at the Southwest of Greenland that reached there flowing with the Arctic stream.

Fridtjof Nanson (Fig. 4.9) was one of the most innovative and adventurous Arctic explorers. He was a Norwegian multifaceted personality. He was an Oceanographer, Zoologist, Nobel Peace Prize Winner (1922), and an eminent personality of Norway. He was a dreamer with the ability to convert his dreams into reality.

In 1888, he made a memorable journey across Greenland on skis. He described this journey in his book *First Crossing of Greenland* (1890). He made his first trip to the Arctic on a sealer in 1882 and upon his return became curator of the natural history collection of the Bergen Museum. He observed that the natives of Greenland, the Eskimos, use the wood of the Siberian forests. After investigations, he found that the woods flow through the rivers of

FIGURE 4.9 Fridtjof Nansen. *Source: https://en.wikipedia.org/wiki/Fridtjof_Nansen#/media/ File:Fridtjof_Nansen_LOC_03377u-3.jpg.*

Siberia and reach the Arctic Ocean and after that they reach the Costs of Greenland through the Southwest stream of the Arctic Ocean. Nansen has also read that the remains of the American ship "*Zenith*," which sank on the coasts of Siberia, were found buried under ice on the Southwest coasts of Greenland.

Based on the above facts, Nansen assumed that flowing toward the west, a stream of the Arctic Ocean might have joined the Atlantic Ocean and the woods of Siberian forests and the remains of the *Zenith* might have reached Greenland with the stream. He also predicted that starting from the North the stream might have reached Greenland crossing through the North Pole and Franz Joseph Island and with the help of this route he could reach the North Pole. Till that time, the North Pole was not conquered.

In fact, this idea of Nansen was quite adventurous, and some people compare his idea with Maglane's assumption of the 16th century about the location of the Pacific Ocean on the other end of South America. Till that time nobody was aware of this. Nansen proceeded ahead to realize his long-cherished dream—North Pole Conquest. He made concrete efforts to experiment with his dream. To fulfill this dream, his first step was to make a strong and durable ship that can face the challenges and come out of the hard ice. With incomparable hard work and brilliance, he was able to make the ship "*Fram*" (which literally means progressive). It was 128 ft high weighing 328 tons. The ship can travel in the polar region.

On June 24, Nansen proceeded toward East on "*Fram*" with his 12 companions who were expert sailors and oceanographers. "*Fram*" sailed forward escaping and cutting the floe ice. Its cabins were very comfortable. For uninterrupted power supply, there was provision of generators who were operated by engines and wind energy. The ship "*Fram*" had an excellent library and was equipped with all the required tools. It had enough capacity to store food for 5 years. Eminent Oceanographer Auto Sverdrup was also accompanying Nansen in "*Fram*." "Fram" followed the same Northeast route through Barents Ocean through which "*Vega*" had successfully concluded its expedition. It had faced a lot of difficulties due to ice floes and the problems intensified in absence of the correct map of the passage. Till the commencement of New Year 1894, "*Fram*" kept sailing and came out of the continental shelf. Nansen found that Arctic is not a shallow ocean rather a deep sea. At one place it was 2850 feet deep. They felt that their assumption about the stream was not accurate, but it was also a fact that the woods from Siberian forests reached Greenland by flowing through the stream. The time kept fleeting. The seasons kept changing. "I" kept sailing on through large ice packs escaping and breaking the ice floes consistently. After 18 months, the day came when Nansen came out of "*Fram*" and decided to reach the North Pole with a Sledge. By that time, traveling Northwest "*Fram*" had crossed 84 degrees north latitude and it was only 580 km away from the North Pole.

On March 14, 1895, Nansen came out of "*Fram*" and proceeded toward the North Pole on Sledge pulled by dogs, with one of his companion Fredrik Johnson. Other companions stayed back in "*Fram*." Nansen's companions were expecting that during summers "*Fram*" will smoothly sail through the stream and reach the coast of Norway. At that time, "*Fram*" was stuck in an ice floe and there was no hope of it coming out of it before summers. Nansen and Johnson were moving with zeal and enthusiasm to conquer the North Pole, but they were facing acute hardships on their way. It was becoming difficult to manage the three sledges. On their way, they had to kill a dog to feed the other dogs. Though it was March, but the ice had melted, and it was impossible to travel by Sledge on water. The journey became cumbersome and dull. Moving slowly and making the camp at the end of the day and sleeping in bags with wet clothes had become their routine. Despite the struggles, the progress was not up to the expectations. Very soon, Nansen gave in to the circumstances. They realized that they cannot reach the North Pole though they had reached at 86°13' north latitude and the North Pole was only 361 km away. No one had reached that point, but it was decided that they will not be able to reach the North Pole.

At last, they had to return. But they did not proceed toward the direction of "*Fram*," instead they moved toward Franz Joseph Island. On the way, their foodstuff got over and they had to kill seals and even their dogs. Somehow when they reached Franz Joseph Islands, winters had arrived, and they had to spend the winters on Franz Joseph Island. They came out of the winter shelter

FIGURE 4.10 Robert Edwin Peary. *Source: https://en.wikipedia.org/wiki/Robert_Peary#/media/File:Robert_Peary_self-portrait,_1909.jpg.*

in next May and proceeded toward South. A miracle took place on 17th June of 1896 when Nansen and Johnson met British explorer Fredrik Jackson. They had heard a lot about expeditions of Jackson before embarking on their journey. Both greeted each other and returned to their respective countries. Hence, Nansen could not realize his long-cherished dream but explored and made a lot of investigations. At last on August 12, 1896, "*Fram*" also returned intact. Various expeditions from many countries were sent to reach the North Pole through Franz Joseph Island after the expedition of Jackson but none of them succeeded in their missions.

11. Robert E. Peary

Meanwhile, Robert E. Peary (Fig. 4.10) appeared on the scene. He traveled twice to the North Pole through Northwest Greenland. In 1900, he explored the North end of Greenland that became popular as "Peary Land." Peary Land generally remains free of ice.

Before this, he wanted to begin his journey to the North Pole from North Greenland but later he considered North Ellsmere Island more suitable for it. During 1898−1902, he traveled along the North Coast of Greenland by Sledge and reached 84°17′ north latitude. In 1905, he again started for the North Pole. This time, Bartle (Capt. Bob) an expert sailor accompanied him. They reached Cape Sheridan and upto 87°6′ in their ship "*Theodore Roosevelt*." This time they also traveled further to the North Coast of Ellsmere Island, but they could not succeed to reach the North Pole this time again. After many unsuccessful attempts, Robert E. Peary undertook their eighth expedition in 1908 to reach the North Pole and this time he was successful, and Peary became the first man to conquest the North Pole. Let us learn about this journey in detail.

From the very beginning, Peary was confident that only he can conquer the North Pole. Determined and totally committed toward his mission, Peary was different from other explorers. Eminent explorers like Terry Hudson and John Davis dreaded the winters but Peary preferred winters for expeditions to the North Pole. The explorers of the first generations were brave enough to spend the extreme winters in Arctic but preferred to travel during summers. Peary had a different vision. In fact, the ice becomes hard enough to travel smoothly and safely. The objective of Nansen's expedition was to collect scientific information about the Arctic, but Peary wanted to reach the North Pole at the earliest. Robert Peary favored the use of dogs and the clothing of Eskimos. He preferred to travel light, but he used to carry enough foodstuff for his companions, including the dogs of the sledge.

It will not be an exaggeration to repeat here that the only mission of Peary was to conquer the North Pole. He never thought of anything else but to reach the North Pole. He believed the explorers who could not succeed in their missions to reach the North Pole had not focused on their objective properly. Peary began his last expedition to North Pole in "*Theodore Roosevelt*" from New York in 1908. After a few weeks, the ship "*Theodore Roosevelt*" was sailing along the West coasts of Greenland. From there, Peary took 150 kg of meat of Walrus and Whale, 246 furious dogs, and 49 energetic Eskimos including some females and 10 children. Crossing through the ice-packed passage between Greenland and Elsmere Island, the ship reached Cape Sheridan on 5th September covering 560 km. This place is 800 km away from the North Pole. Peary was very popular among Eskimos. He had become the best friend of the Eskimos of Smith Sound located on the North West coast of Greenland. They used to provide him all the necessary things for his expedition such as sledge, dogs, food, and hunters. Peary had great admiration for the tolerance and courage of the Eskimos in overcoming the adversities with ease and a smile. He wrote in his diary "I love the childlike innocence of these people. I know the name of each member of the group. I saved them from starvation."

The elders of the troop used to advise their young children "If you grow as a good hunter, Peary will award you." During his journey to Cape Sheridan, the Eskimo males kept making sledges and females stitched clothes for the members of the expedition on the ship. The evening started with September in the Polar region. The hunters of the ship started hunting the Caribou and Musk Ox in the interior regions and soon enough food was arranged for onward journey. The journey started for the next camp at Cape Columbia 145 km away. After spending the winters at Cape Columbia, in the beginning of Sprint,

Peary moved forward to the last camp at the North Pole. They left Columbia and started a further journey by sledge. The expedition started in February 1909. The six members of this last part of the expedition were viz. Robert Barlet (Capt. Bob), Surgeon Dr. J W Gudsale, Secretary of Peary, Rous Marwin, two young investigators George Burrup and Donald McMillan, and a Negro friend Mathew Henson. For necessary assistance, 17 Eskimos and 133 dogs were also included in the expedition. They were carrying luggage of 3250 kg and the North Pole was 800 km away. Later, Peary wrote in his book "North Pole" about this journey "It was my policy to take minimum work from the assisting team. It was essential to keep the core group of this last expedition hale and healthy till last minute. Though I had selected the group in the beginning of the expedition. I made all efforts to make the journey comfortable in every possible way."

The last part of the journey was very troublesome. It was essential to cover the predetermined distance every day. There was the least time to sleep and rest. Apart from facing the heavy winds, they had to cross the high ice rocks. The most challenging part of the journey was the leads of water due to the melting of ice. They had to wait for the refreezing of the ice so that they can continue back on their expedition. On one occasion, the expedition group had to wait for 6 days for refreezing of the lead. Facing and overcoming the obstacles with determination and confidence, the group reached a place that was just 135 nautical miles from the North Pole. Peary adopted platoon methodology for the expedition that is used to climb the high peaks. They used to leave behind the less useful dogs and members of group. On April 1, 1909, Peary sent back Capt. Bob and assisting members of the group to South and proceeded toward the North Pole again. Now, he had with him best dogs, three experienced Eskimos, his best companion for the past 20 years, 43 years old Mathew Henson, the Negro. Apart from being an expert investigator, Henson was an excellent sledge driver. The weather was very pleasant on April 1, 1909.

Peary prayed to God to continue this pleasant weather for another 3 days and the God listened to his prayer and fulfilled his wish. The weather remained pleasant for another few days. The weather was nice, and the path was quite smooth with a thick hard layer of ice on it. The goal was quite near. Therefore, Peary moved ahead with speed. They fixed the goal of traveling 40 km per day and achieved it, though most of the time they traveled by sledge. The front part of their feet had decayed during their journey to Elsmere Island in 1899. After traveling for 6 days, on April 6, 1909, Peary measured the Altitude at 10 a.m. and jumped with joy. They had reached $90°57'$ and the North Pole (90 degrees) north latitude was just 3 km away. The goal for which all were struggling for the past 20 years was in front of them. They controlled their emotions and moved forward toward the North Pole. After few moments, Peary and his friends were standing on the North Pole (Fig. 4.11). They had conquered the

FIGURE 4.11 Robert Peary Sledge Party and flags at the Pole. *Source: https://upload.wikimedia. org/wikipedia/commons/9/95/Peary_Sledge_Party_and_Flags_at_the_Pole.jpg.*

North Pole. Despite the enigmatic mental and physical condition, he wrote in his diary "I was tired. I was not able to believe that I have attained my goal."

Mathew Hansen (Fig. 4.12) was the most reliable companion of Peary who accompanied him in all eight expeditions. Son of a farmer, he started working in the ship from the age of 12 only. He has narrated the North Pole Conquest in a very interesting way in his book "A Negro at the North Pole." Subsequently Hansen was also conferred an honorary degree by the Harvard University.

Peary then reached into his over garment and took out a rolled-up flag seamed by his partner and fixed it to an employee, that he stuck atop Associate in Nursing hut that his native companions had designed [3]. After that, he hoisted four more flags of American Navy, Fraternity League, D.A.R. (Peace), and Red Cross. All the five members took photographs with the flags; cheered and shook hands with each other. In addition to this, Peary did a unique thing that was quite uncommon for the place and occasion. He was very tired, and his body was in need of sleep. He slept for another 3 h. For another 30 h, the members of the expedition undertook sounding experiments on the ice and found that the ocean was more than 1500 feet deep under the North Pole. They made observations on the weather and climate. After that, the crew moved toward South (at the North Pole, all the directions lead to South). The return journey was quite comfortable as scheduled. On April 25, 1909, Peary returned to "*Roosevelt*" and the ship proceeded toward America.

FIGURE 4.12 Mathew Hansen. *Source: https://en.wikipedia.org/wiki/Matthew_Henson#/media/ File:Matthew_Henson_1910.jpg.*

12. Challenge to Peary

Before Peary could return to the United States of America in September 1909 Frederick A Cook announced that he had reached the North Pole on April 21, 1908, a year before Peary did the expedition. Cook (Fig. 4.13) was one of the members of Peary's group during his expeditions of 1891–92 to Greenland. He was in the Arctic during 1907–09.

Cook was the son of a German tribal immigrant to America. He had accompanied Amundsen as a doctor during his journey to the Arctic. According to Amundsen "Cook was a brave but a cunning person." Cook's claim generated a serious controversy on the North Pole expedition. To justify his claim Cook produced evidence in Copenhagen University but they were not proved sufficient. The National Geographic Society that had supported Peary in his expeditions to the North Pole advised Herbert an eminent explorer, to investigate the dispute. During his investigations, Herbert found that the Chronometer (a clock to know the exact geographic location) of Peary was 10 min faster. Herbert concluded that the route adopted by Peary was 30–60 miles (48–96 km) in the West of the North Pole. After that, in a study conducted by the National Geographic Society in 1989, it was concluded that Peary's last camp was within 8 km of the North Pole. The study was conducted by a Commercial Navigation Society of the United States.

The recognition of the North Pole Conquest could not be given to any of the two explorers due to a lack of consensus. But the Congress of the United

FIGURE 4.13 Frederick A Cook. *Source: https://en.wikipedia.org/wiki/Frederick_Cook#/media/ File:Frederick_Cook_%D1%81._1906.jpg.*

States and other geographic agencies of the world recognize Robert E. Peary as the conqueror of the North Pole. A brief life sketch of Robert E. Peary is being given. Robert Edwin Peary was born in Cresson, Pennsylvania, on May 6, 1856. He graduated from Bowdoin College in 1877 with the degree of Civil Engineer, standing second highest in his class. In his early career years, he worked as a draftsman in the Coast and Geodetic Survey offices in Washington, DC, and he entered the Naval service as a Civil Engineer, USN, with the rank of Lieutenant, on October 26, 1881. He worked as Sub-Chief with the Interocean Canal Survey in Nicaragua. The man, who spent most of his life in a cold region like the Arctic, started his career in a tropical region. During his Nicaraguan Canal assignment, he became interested in the far north. The attraction was so irresistible that North Pole exploration became his aim and objective of his life. In April 1886 he took 6 months leave of absence to lead an expedition to Greenland where he recorded important ethnological and meteorological observations. During this journey, he was accompanied by his wife Josephine Diebitsch and their children. He wrote in his book "in such journeys women prove to be a meaningful companion as they are equally effective in observation."

After working for a short period in Nicaragua, he returned to the Arctic again in 1891—92. This time, he traveled 2100 km from Mac Cormack Bay to Independence Bay in Greenland. During this journey, he found out that Greenland is actually an Island. He undertook brief journeys of Cape York and

Greenland to find the remains of meteors explored earlier. The first attempt to reach the North Pole was carried out over a period of 4 years, from 1898 to 1902, when a point 343 miles from the pole was reached. They could reach upto 84,017′ north latitude. During their seventh expedition, they reached upto 8706′. The North Pole was only 322 miles from that point. At last on April 6, 1906, they reached the North Pole during their eighth expedition. During his last expedition, Peary was in the commission of Yankee Navy and thus he hoisted the flag of Navy on the pole that was placed in a very little glass bottle and born into a crevice within the ice: 'I have these days hoisted the national ensign of the United States at this place, that my observations illustrate to be the North Polar axis of the planet, and have formally taken possession of the complete region, and adjacent, for and within the name of the President of the United States [4].

Some people believe that it happened because of his friend Theodore Roosevelt, the then President of the United States. Peary retired as Rear Admiral. He spent the rest of his life providing flight training to people. He died in Washington, DC, on February 20, 1920, and is tombed in Arlington National Cemetery.

13. Nautilus under the North Pole

The French Novelist Jules Verne was fond of writing science fiction. Based on his scientific knowledge and imagination, he wrote many novels that became immortal forever. He imagined crazy Americans who took off from Florida in a specially designed spacecraft to travel around Moon and returned successfully. In his novel "20000 Leagues under the Sea," he imagined a submarine that traveled 65,000 miles under the sea level. His first imagination was realized by the Apollo-10 and his second thought was realized into *Nautilus*, the submarine of the United States. The name of Jules Verne's imagined submarine was also Nautilus and it was operated by hydroelectricity. But the real *Nautilus* was operated by atom energy. The captain of Jules Verne's Nautilus was Nemo and the Captain of real *Nautilus* was Anderson.

The plan of exploring under the North Pole through submarine was not new. George Hubert Wilkins had also tried to execute his plan in 1931, but his submarine was powered by a diesel engine and its operating devices were not suitable for the polar climate conditions. Therefore, the efforts of Wilkins could not be successful.

Though it was not easy for an atomic-powered submarine to travel continuously for thousands of miles as navigation beneath the arctic ice sheet was difficult. But, it was essential to devise a new system before such expeditions. Therefore, *Nautilus* (Fig. 4.14) was powered by the S2W naval reactor, a pressurized water reactor produced for the US Navy by Westinghouse Electric Corporation and Bettis Atomic Power Laboratory. Nuclear power has a crucial advantage in submarine propulsion, because it is a zero-emission

FIGURE 4.14 Nautilus underway. *Source: https://en.wikipedia.org/wiki/Nautilus_(fictional_ submarine)#/media/File:Nautilus_Neuville.JPG.*

process that consumes no air. Above 85°N, both magnetic compasses and normal gyrocompasses become inaccurate. A special compass engineered by artificer Rand was put in shortly before the journey [5]. The submarine *Nautilus* had all facilities for comfort, entertainment, and knowledge development. It had a large library of films, large high tech Mess, Coca Cola Machine, and a Jukebox.

Equipped with all safety measures, *Nautilus* set off toward the ice packs between Greenland and Spitsbergen on August 19, 1957, under a secret mission. Its goal was to reach near 83 degrees north latitude. In other words, it had to travel 385 km under the ice. The journey was successful with a slight defect in the scrubber of carbon dioxide. After a couple of trials, Captain Anderson set off *Nautilus* toward the North Pole at 8 pm on September 1, 1957. Captain Anderson directed the submarine toward North. Its goal was to go deep inside the Arctic and if possible upto the North Pole, which was 1050 km away. *Nautilus* was about to reach 86 degrees north latitude, when it was observed that both magnetic compasses and normal gyrocompasses became inaccurate. Despite this, Anderson tried to proceed ahead but after crossing 87 degrees north latitude it became clear that moving ahead will invite the trouble of "longitudinal Roulette." Longitude Roulette is a condition when the magnetic compasses become disoriented and start showing wrong directions and rather going deep into the sea, the ship moves toward shallow waters. The *Nautilus* had to return and was taken over by NATO.

When *Nautilus* was under war exercise for NATO, the strategy for the North Pole journey was also being prepared. This time it had to commence its journey from the Pacific Ocean. Passing through Bering Passage, it had to enter the Arctic Ocean by moving toward the West to reach the Atlantic Ocean. It also had to pass through the North Pole as well. To execute this plan, *Nautilus* began its journey from Aiho (Japan) as per the predecided route on July 25, 1958. On its journey, it passed through more than 1 km long, thick ice caps but it kept moving at an average speed of 19.6 naught continuously for 6 days and reached the Arctic Sea. In the Arctic it had to travel deep under the sea to avoid the large ice caps, sometimes the distance from the seal level was 8 m only. Observing the large ice caps, their features and expansion, the large mountains, the planes and trenches of under the Arctic, *Nautilus* kept moving forward. Passing through "Pole of Inaccessibility" at 83°20′, it reached 87 degrees north latitude. "Pole of Inaccessibility" is a geographic area of the Arctic that is full of ice caps considered to be the most inaccessible region. Now, *Nautilus* had to pass through the Arctic ice. Though the Arctic has a thick layer of ice with extreme cold winds, the environment inside the *Nautilus* was very comfortable and warm. Traveling for 62 h, it reached at a place when all the crew members became restless and excited. The moment to create the history had arrived. The North Pole was merely 0.6 km away. Anderson could not stop himself and said, "We should pray for the universal peace and thank all those people who have been supportive and helpful in making our expedition a success."

On August 3, 1958, at 23:15 h, the US nuclear-powered submarine Nautilus accomplished the primary submarine voyage to the geographic North Pole [6]. As per the local tropical time, it should have been dark at that time, but the midnight sun was shining above them. On reaching the North Pole, the crew members cut the specially designed cake "Panopto" (Pacific to Atlantic via the North Pole). Anderson took a sample of the ice at the North Pole. The temperature of the water at the North Pole was measured as 0°C (32.4°F) and the depth of the sea was 2235 Fathom. This depth was more than what was measured by Evan Papanin in 1937 and Peary in 1909. The ice just below the North Pole was 8 m thick. In this way, *Nautilus* proved that 7850 km can be saved in a journey from Japan to Europe. Now moving toward Europe, *Nautilus* had to inform Washington about the success of the expedition. On its return journey, it had to face and pass through 19 km long ice caps. The water was getting colder and the depth of the sea got 2400 fathom deeper. The fear of Longitude Roulette kept on prevailing throughout the journey. But *Nautilus*

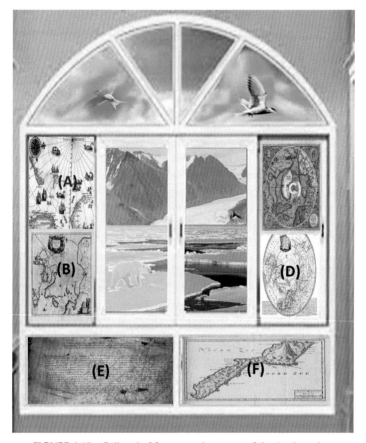

FIGURE 4.15 Collaged of famous ancient maps of the Arctic region.

overcame all the obstacles. At last, after continuously traveling 2925 km for 96 h, it came across a hole on the North East Coast of Greenland and it came on the seabed. The message "Nautilus 90 North" was sent from there, which meant Nautilus has reached 90 degrees north latitude safely. Then, submarine *Nautilus* proceeded toward Europe, but a helicopter was waiting for Anderson in Reykjavik, the capital of Iceland, which took him directly to Washington. Anderson was felicitated in Washington and he presented the sample of the ice from the North Pole to Admiral Rickover.

14. Ancient mapping of the Arctic region

Some of the famous ancient maps [7] of the Arctic Region are collaged in Fig. 4.15, which opens a window to the capabilities, wisdom, and cartographic knowledge of these earlier sailor and map engravers.

A. Spitsbergen and Svalbard during the Golden Age of Dutch exploration and discovery (c.1590−1720s). A portion of 1599 map of Arctic exploration by Willem Barentsz. Spitsbergen, here mapped for the first time, is indicated as "Het Nieuwe Land" (Dutch for "the New Land"), center-left. This is a typical map from the Golden Age of Netherlandish cartography Jan Jansson's map of the "Poli Arctici" from 1644.

B. Jan Jansson's map of the "Poli Arctici" from 1644.

C. Mercator's map of the North Pole (1606)

D. Emanuel Bowen's 1780s map of the Arctic features a "Northern Ocean."

E. Patent from King Henry VII, authorizing John Cabot and his sons to explore new lands in the west

F. A Dutch map of Jan Mayen during the Golden Age of Dutch exploration and discovery (c.1590−1720s). The Dutch were the first to undisputedly explore and chart coastlines of Jan Mayen and the Svalbard archipelago in the Arctic Ocean.

References

[1]   https://en.wikipedia.org/wiki/Major_explorations_after_the_Age_of_Discovery.
[2]   The Nautilus Expedition, November 20, 1931. Amphilsoc.org. Retrieved 8 July 2014.
[3]   https://www.smithsonianmag.com/history/who-discovered-the-north-pole-116633746/.
[4]   https://www.teaparty.org/putins-push-arctic-ocean-53860/).
[5]   https://en.m.wikipedia.org/wiki/USS_Nautilus_(SSN-571).
[6]   https://homeschoolontherange.blogspot.com/2018/11/.
[7]   https://en.wikipedia.org/wiki/Arctic_exploration.

Chapter 5

The society and the living style of its natives

During the Wisconsin glaciation (c.50,000−17,000 years ago) due to falling sea levels people could cross the Bering land bridge between Siberia and northwestern/North America (Alaska), which yielded human habitation in North American polar region [1]. Subsequently, diversified sects of ancient Eskimo people emerged and spread across Arctic North America, such as Pre-Dorset (c.3200−850 BCE), the Saqqaq culture of Greenland (2500−800 BCE), the Independence I and Independence II cultures of northeastern Canada and Greenland (c.2400−1800 BCE and c.800−1 BCE), the Groswater of Labrador and Nunavik, and the Dorset culture (500 BCE to 1500 CE). Perhaps, the Dorset were the last major Paleo-Eskimo culture in the Arctic. Thule, the ancestors of the modern Inuit [2], lived from about 200 BCE to 1600 AD, around the Bering Strait. These Thule people are still living in Northern Labrador [3].

1. Indigenous people of the Arctic

The Arctic is home to about four million individuals, together with autochthonous peoples. The diversified style of language, cultures, and traditions provide a unique identity to them. Accordingly based on the culture and the language of the indigenous people, the communities are divided into various categories (families) (Figs. 5.1A and B). Indigenous peoples are simply tormented by environmental modification and enlarged economic activity within the Arctic. One must examine how we can contribute to achieve sustainable development of that autochthonous peoples can see advantages while protective the foundations of ancient cultures and lifestyles. Though the Arctic has become a vital issue to the international community in recent years, scientific understanding of the Arctic continues to be inadequate. The changes within the Arctic and their influence on the world as an entire should be understood with a comprehensive and wide-ranging perspective, considering the climate, material cycles, diverseness, and therefore the effects of human activities. It is necessary to clarify the mechanisms or causes of the changes and to predict the changes occurring in the future, and to strengthen comprehensive studies for assessing those socioeconomic impacts.

The Arctic. https://doi.org/10.1016/B978-0-12-823735-9.00016-3
69

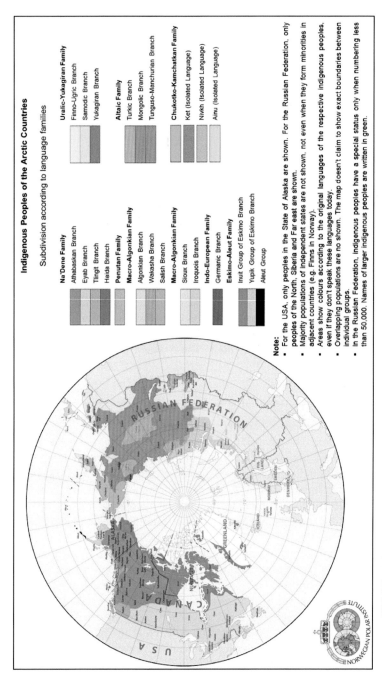

FIGURE 5.1A Familywise classification of Arctic indigenous peoples. *Source: https://ansipra.npolar.no/image/Arctic04E.jpg.*

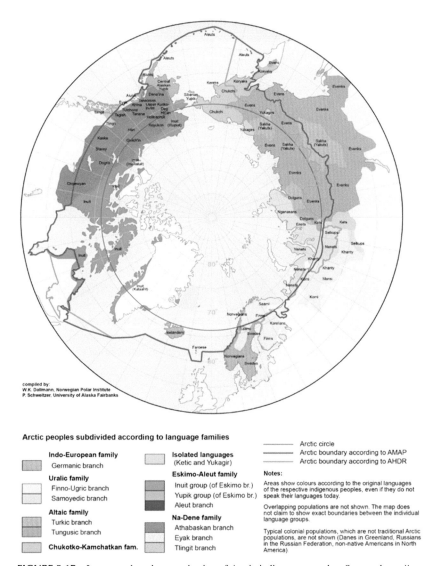

Arctic peoples subdivided according to language families

Indo-European family
Germanic branch

Uralic family
Finno-Ugric branch
Samoyedic branch

Altaic family
Turkic branch
Tungusic branch

Chukotko-Kamchatkan fam.

Isolated languages
(Ketic and Yukagir)

Eskimo-Aleut family
Inuit group (of Eskimo br.)
Yupik group (of Eskimo br.)
Aleut branch

Na-Dene family
Athabaskan branch
Eyak branch
Tlingit branch

———— Arctic circle
———— Arctic boundary according to AMAP
———— Arctic boundary according to AHDR

Notes:

Areas show colours according to the original languages of the respective indigenous peoples, even if they do not speak their languages today.

Overlapping populations are not shown. The map does not claim to show exact boundaries between the individual language groups.

Typical colonial populations, which are not traditional Arctic populations, are not shown (Danes in Greenland, Russians in the Russian Federation, non-native Americans in North America)

FIGURE 5.1B Language-based categorization of Arctic indigenous peoples. *Source: https://epr. arctic-council.org/images/PDF_attachments/Maps/languages.pdf.*

Further, as the Arctic warms, attracting interest to the region, the autochthonous communities face new challenges. Thawing soil is inflicting damages to infrastructure because the ground becomes less stable. Developing new infrastructure to support economic development would force innovative approaches within the region not knowledgeable about earlier. The supply connected difficulties such as transporting building materials and experience can more compound the difficulty. The warming trends, however, will modify more economic activity within the region. Decreasing ocean ice coverage is sanctioning larger maritime traffic.

2. Alaska

In the year 2003, the military business firm of Engineers was knowing that four Alaskan villages would want to relocate owing to the risk of flooding and erosion. There are other additional villages in the list from the time frame between 2003 and 2016. One among these villages, Shishmaref has a population of 650 people who are exposed to the danger of ongoing climate change. However, relocation is tough because of no governmental institutional framework that exists for the help of climate refugees within the United States.

3. Sami people

People primarily from north Sweden, Norway, Finland, and Russia who survived for generations by fishing and hunting are known as Sami people. The Sami people have come to rely on and predict the changes in climate causing unpredictable conditions in the ice so that they and their herds can move across it. Once 300 reindeers sank in the ice and drowned. Land grabbed by the Swedish government is also prevalent and lacks in communication between them and the Sami people.

4. Inuit

The Inuit are a group of culturally similar indigenous peoples inhabiting the Arctic regions Greenland, Canada, and Alaska (Figs. 5.2A−C).

5. Displacement

In the Arctic region, phylogeny activity has been increased attributing to the resource race. This led to the threat of displacement of habitant's endemic to the region because the climate changes within the region and native animals' traditional patterns square measure discontinuous. accrued phylogeny activities square measure poignant the communities' food supplies; melting land and erosion has broken native infrastructure, as well as homes, buildings, waste product systems, etc. Unstable and unpredictable ice patterns have affected quality important for transportation, hunting, travel, and communication. Moreover, the changes and variability in the climate have left indigenous communities who rely on traditional knowledge vulnerable essentially "strip arctic residents of their considerable knowledge, predictive ability, and self-confidence in making a living from their resources."

FIGURE 5.2A Women's traditional caribou skin outfit with *amauti* parka, trousers, mitts, and footed stockings. The back of the parka has an *amaut* or pouch for carrying a baby *Source: https://en.wikipedia.org/wiki/Inuit_clothing#/media/File:T%C3%B8j_til_kvinde_fra_Rensdyr-inuit_i_arktisk_Canada_-_Woman%E2%80%99s_clothing_from_Caribou_Inuit_in_Arctic_Canada_(15307253096).jpg.*

6. Environmental degradation

6.1 Economic implications

While some indigenous communities believe that the resource race will provide economic opportunity, others are skeptical of how much it will benefit them economically. In the past, economic benefits of resource extraction in the Arctic have created revenue for governments and private entities, while relatively little if any of its wealth was directly returned to improve the economic and social well-being of the local people in the regions where extraction occurred.

6.2 Food security

The dynamic climate within the Arctic has effects on food security among endemic communities. Once the ice is unsuitable to travel on, it is not possible to seek food. Thus, food security is prone to global climate change because the food offer consists of native species that square measure themselves sensitive

FIGURE 5.2B Man's parka and pants, southern Baffin Island Inuit, Hudson Bay (1910–14)—Royal Ontario Museum. *Source: https://en.wikipedia.org/wiki/Inuit_clothing#/media/File:Man's_parka_and_pants,_Inuit,_southern_Baffin_Island,_Hudson_Bay,_1910-1914_-_Royal_Ontario_Museum_-_DSC00302.JPG.*

FIGURE 5.2C Modern Inuit women in traditionally constructed amauti; left: seal, right: caribou. *Source: https://en.wikipedia.org/wiki/Inuit_clothing#/media/File:Inuit-Kleidung_1.jpg.*

to environmental condition changes. There is conjointly concern over the toxins found in native species and therefore the risk of oil contamination in endemic communities' food provides. The full population of the Arctic region is 40 lacs and most of the folks are not the natives of the Arctic. The essential natives measure solely 100% of the full population. These residents conjointly coming back involved with the fashionable world and losing their basic life-style, rituals, and culture.

The Arctic region includes countries like Norway, Sweden, Finland, Scandinavian country (Greenland), Iceland, Russia, Canada, and Northern components of the United States. The residents of those countries accustomed visit the Arctic in search of Minerals and alternative natural resources and settle there once it slow. In the 20th century, an oversized range of population from the higher than countries settled in the Arctic. Like others, they are living in cities and have started several professions for survival. On the alternative the opposite hand owing to the migrants from other regions, land disputes square measure arising. So just like the Noonavat of Canada, efforts square measure being created to produce political and economic autonomy to the region for the protection of original natives. There is a special lifestyle of the initial natives of the Arctic that is not in tandem with the fashionable civilization. They need distinctive life designs, rituals, superstitions, and beliefs. The word Eskimo (Fig. 5.3) visualizes a stout person with fierce dogs pulling the sledge wearing leather from head to toes.

We also see his house Igloo (Fig. 5.4). The Eskimos are unique and may be more unique than our imagination. One thing to be pointed out here is that Eskimos do not like to be called as Eskimos. They call themselves Inuit. Inuit means public. Eskimo word has been derived by Red Indians, and the meaning given to it by the Europeans is "one who eats raw meat." This definition given to this Eskimo is wrong, and no one would like to be called by this word. Despite refusing this word Eskimo by the natives, most of the people from other countries called them as Eskimos. They are defined as Eskimos in different books and dictionaries. Therefore, we would use this word in our further discussions.

7. Unique qualities

Surviving in harsh climatic conditions of Northern parts of North America, Europe, and Asia and Islands of Arctic Ocean, Eskimos are people with unique qualities. Their life is full of controversies. They spend most of their life in dangerous conditions. These dangers are the fierce ice storms, sinking of smooth ice, and bad weather (most of the time the weather is difficult). They live so near to death that they do not want to take the risk of looking after their elder ones. The elder members accept the death and do not want to be a burden on their families. The dogs are an integral part of their life and are very useful

FIGURE5.3 Eskimo. *Source: https://en.wikipedia.org/wiki/Eskimo#/media/File:Koryak_armor.jpeg.*

FIGURE 5.4 Igloo. *Source: https://upload.wikimedia.org/wikipedia/commons/7/75/Igloos.jpg.*

for them. They cannot hunt without them especially during winters; however, when there is nothing to eat during extreme climatic conditions, they kill the dogs for food. According to some people, they even kill human beings when such conditions arise.

Despite all this, the Eskimos are happy, motivated, and courageous people who always keep smiling. During extreme climatic conditions, when there is nothing to eat, they entertain the guests with delicious food and try to keep them happy. Despite being aware about the dangers of travel, the Eskimo families keep traveling to faraway places on sledge. The Arctic is not suitable at all for agriculture, and there are no trees with fruits. Therefore, the life of the Eskimos is dependent on hunting of Walrus, Whale, Reindeer, and Seal, which is a difficult task in itself. Even when the Eskimos cannot get anything for food by hunting even then the Eskimos do not enter their Igloos in frustration. He keeps smiling even at that time. Despite built of Ice, the Igloos are always warm from inside. In fact, the Igloo is so warm that one has to take off the outer clothes to avoid sweating.

From Greenland to Baring Sea approximately 50,000 Eskimos reside in the Arctic region. Though their population is not equal everywhere, their language is the same. Based on this and other cultural qualities of Eskimos, the Anthropologists have assumed that Eskimos emigrated in the areas of Alaska, Canada, and Greenland approximately 2000 years back. Their origin is not known yet. The Anthropologists are of the view that the human population is 10,000 years old. These ancient natives developed their own culture that was adopted by the Eskimos. As per the historians, there is a mention about the Vikings during the 11th century who settled in Greenland, those were Eskimos only. The history is the witness that after the Ice age, due to the freezing of the climate, many habitats and species were lost. But during the 16th century, the Europeans met the Eskimos for the trading of whale, seal, and other fur animals. At that time, Eskimos were not aware of any metal. They got the guns and other arms from Europeans and started hunting with them. Though till date they make sharp-edged arms and knives from the bones of whales, Walrus, and Caribou.

8. The expert inventers

Eskimos are exemplary of survival under extreme conditions and winners. Due to the scarcity of wood in their environment, they used ice to make their houses. They used the fat of seal for lightning their lamp and invented sledge to travel on the ice. To satiate their hunger due to the impossibility of agriculture, they used meat. Made arms from the bones of animals due to lack of knowledge about metals. In the absence of cotton, silk, or wool, they made clothes from the skins of animals and used the needle and thread made from the bone and organs of the seal.

In this way, they made many inventions to overcome the conditions. Therefore, Eskimos are called unique inventors. Though according to some historians these inventions were made by their ancestors, but the Eskimos made necessary additions and improvements in them. Some historians say that most of the inventions were made 20,000 years back during the Stone Age by the natives of middle Asia and Europe. The Igloos were invented by Eskimos.

9. Family life

The needs of Eskimo tribes are very limited. Therefore, their way of living is very simple. Their main necessity is search for food and they relish it with their family. The hunting for food is done by males (Fig. 5.5) but the cooking, cleaning of house, and stitching of clothes are done by the females. The boys fulfill the responsibilities of their father and the girls of their mothers like other tribes. The boys must make the arms, bows and arrows, harp, etc. for hunting, taking care of the dogs for the sledge, and making the kayak (a boat made of leather). He must learn to make the Igloo. The girls must prepare to share the load of the male counterparts apart from learning the household chores of their mother.

Human Rights advocates demand that the Inuit and other indigenous people from the area are consulted in any development process. But there are practical skills in survival that must be retained for the good of everyone. Signatories to the United Nations Convention on Biological Diversity are under an obligation to respect the traditional ways of life. Approximately four million people are living in the Arctic, 10% of which are indigenous people. In the Arctic Council, indigenous people have permanent representation. Most of the indigenous communities' fear resource exploitation resulting in negative environmental impacts that will negatively affect their wellbeing. For some people, it is an important economic opportunity for those who are struggling to adapt to changes in the regional climate.

Accelerated climate change in the Arctic as a result of resource exploitation and increased anthropogenic activity in the region will drastically alter the livelihoods of indigenous people in the Arctic. Indigenous people depend

FIGURE 5.5 Seal hunter waiting. *Source: https://en.wikipedia.org/wiki/Seal_hunting#/media/File:Seal-skinning_on_the_floes.jpg.*

entirely on the natural environment for their necessities such as hunting, harvesting, fishing, and herding. Melting sea ice and extreme weather patterns threaten the animals that survive off the established conditions and thus threaten the people that depend on such animals for food. Changes in this natural environment will have impacts on their economy, society, culture, and health. Indigenous people have listed their main points of concern viz: contaminants, land use, climate, security, and access.

10. Impact of Arctic climate change on indigenous people

Indigenous populations that have underpopulated the Arctic for thousands of years are seeing Arctic coastlines rapidly erode, resulting in the destruction of buildings and infrastructure, and endangering native communities, notably those built on permafrost. More and more, whole communities must be compelled to relocate to avert a crisis. As of 2013, 12 out of 31 at-risk Alaskan communities planned to maneuver to new locations; however, they long-faced difficulties distinguishing acceptable new sites, building new infrastructure, and funding their moves.

The size of autochthonic communities within the Arctic is important. According to some, autochthonic peoples are composed of half Canada's Arctic and most of the Greenland's population. Moreover, there are concerning 600,000 autochthonic folks in Russia that include around 10% of the country's Arctic population.

Due to past industrial and government activities, the Arctic is additionally host to dangerous pollutants, as well as serious metals, oil and crude oil merchandise, nuclear radiation, and chronic organic pollutants (POPs). These pollutants cause serious risks to the whole organic phenomenon. Free POPs are eaten by Arctic mammals, like beluga whales, ringed seals, and polar bears, and keep in their fatty tissues. This poses a heavy health risk to autochthonic people and other consumers including hunters. Accordingly, cases of autochthonic peoples with high levels of poisons in their blood and breast milk are on the increase.

11. Protection of the environment and its people

The effects of temperature change within the Arctic have globally relevant repercussions viz. witnessing rising ocean levels ensuing from ice loss on the Greenland ice sheet and altered weather patterns caused by the perturbation of jet streams. Such consequences conjointly place native populations at nice risk. Their living and well-being are directly coupled to the land, however, their management over its preservation is uneven, starting from substantial (in North America and Greenland) to restricted (in Fennoscandia and Russia). Semipermanent changes within the Arctic are driven primarily by external factors, like world commodity costs and rising gas emissions. Therefore, future environmental, economic, and social developments within the region rely critically on policy and business choices created elsewhere at the national and international levels, like progress in global climate change negotiations. At present, robust disparities exist among national policies on economic development, aboriginal rights, climate change, and environmental protection. Because the region is tiny and ecologically fragile, such inequalities heighten the chance to all or any stakeholders. For instance, robust protections to forestall oil spills can be enforced in some however not all Arctic regions, resulting in an impact outside the protected areas. Agreeing on common policies and legislation across the Arctic countries is therefore crucial, as is that the adoption of property operational standards by business participants.

References

[1] T. Goebel, M.R. Waters, D.H. O'Rourke, The late pleistocene dispersal of modern humans in the Americas, Science 319 (5869) (2008) 1497–1502, https://doi.org/10.1126/science.1153569. PMID 18339930.

[2] The Prehistory of Greenland Archived 16 May 2008 at the Wayback Machine, Greenland Research Centre, National Museum of Denmark, accessed 14 April 2010.

[3] R.W. Park, Thule Tradition, Arctic Archaeology, Department of Anthropology, University of Waterloo, 2015.

Chapter 6

Flora of the Arctic

A large part of the Arctic, including both land and sea, are covered with snow during a maximum period of the year. So, some people misunderstand that there is no vegetation and habitation in the Arctic. It is true that the different plants and trees that grow in the other regions of earth like tropical and semitropical regions cannot survive and grow in the Arctic as the land is covered by ice. Hence, agriculture is not possible in this region. The vegetation is not grown, it grows naturally, and the plants and vegetation have acclimatized themselves according to the climate of the Arctic. Let us now discuss about the vegetation of Arctic land. Approximately 1700 species of plants carry on the Arctic Champaign, together with flowering plants, dwarf shrubs, herbs, grasses, mosses, and lichens but no trees. Champaign is characterized by soil, a layer of soil, and part rotten organic matter that is frozen year-round. Plants even have tailored to the Arctic Champaign by developing the power to grow beneath a layer of snow, to hold out chemical processes in extraordinarily cold temperatures [1]. In fact, flowering plants have adapted to produce flowers quickly once summer begins. A small leaf structure is another physical adaptation that helps plants in surviving.

1. Vegetation

Permafrost persists in the land areas of the Arctic for a long duration, and there is darkness continuously for 4 months. Due to the absence of sunlight, big trees cannot grow there neither agriculture is possible. Only those plants that can bear the extreme climatic conditions grow there. Some common features are found in all plants. All plants have a small root system. During the summer season only, the surface layer of the ice of the land melts therefore roots cannot penetrate deep into the earth. The leaves of all plants are small so that they can conserve moisture. The plants are small and grow in bunches to keep themselves near the earth so that they can protect themselves from the low temperature and high winds. Most of the plants are perennial. They survive during winters, and this is the reason that Lichens and Mosses are found abundantly. Though during summers in June and July when the ice on the layer of the earth melts, plants with beautiful flowers also grow in the region.

The Arctic. https://doi.org/10.1016/B978-0-12-823735-9.00003-5

1.1 Lichens

Very small plants that look like scum possess many distinct characteristics. According to Botanists, it is a burning example of the symbiotic relationship between algae and fungi. Algae and fungi, both do their work and help each other to grow and survive. Lichens can grow without soil. They can even grow on rocks and thus are called "rock flower."

Its small roots can take nutrients even from very hard substances like rocks. During this process, they decompose chemically and break the rocks, and rocks are finally converted into soil. In the soil formed by such a process, Mosses and Ferns can grow. Later, other plants and shrubs can also grow in this soil. Lichen can also grow on the trunks of trees and can absorb nutrients from them. In this way, it becomes a parasite.

The Arctic is the most favored area of Lichens. More than 15,000 species of Lichens grow in the Arctic region. Some are glued to the trunks of trees and rocks and some look like small bushes. Some like to stick and hang on the plants.

Lichens are found in many colors apart from green. Because of the scarcity of food, many animals depend on Lichens in the Arctic region. A special species of Lichen *Cladonia rangiferina* is the favorite food of Reindeer and Caribou (Wild Reindeer) and thus the name Reindeer Moss. In Scandinavia, it is used for the preparation of bread also. During winters, Caribou feeds on reindeer moss only and without it the Caribou cannot survive. They smell the Lichens even underneath the snow and erode the snow with their hooves to get it. Small mammals of the Arctic region use layers of Lichens to warm their nests or burrow (Figs. 6.1 and 6.2).

FIGURE 6.1 Arctic Lichen. *Source: https://commons.wikimedia.org/wiki/File:Cladonia_rangiferina_205412.jpg.*

FIGURE 6.2 Another variety of Arctic Lichen. *Source: Photo Courtesy Shabnam Choudhary, personal communication.*

1.2 Moss

Mosses do not have roots and flowers. They have only thread like small rhizoids, but small leave grows on it. Rhizoids take moisture and nutrients from the soil. On moist soil, the luxuriant growth of mosses looks like a beautiful carpet floored on the earth. Mosses not only grow in the moist region but can also grow in regions with less soil. They can also grow on hard rocks and take their nutrients from the rocks. Like Lichens, Mosses (Fig. 6.3) are also found in different colors. Small mammals and birds use moss as a lining for their nests.

1.3 Arctic Willow

The famous Arctic Willow (Fig. 6.4) plant looks like a dwarf bush. It is not much high that is why it is able to protect itself from snowstorms. It can grow on soil covered with ice due to its small root system. Sometimes, it creeps on land like carpet.

Eskimos call it "tongue plant" due to its tongue-like appearance of its leaves. Although the color of the leaves of Arctic Willow is green, it turns red during summers. It is the favorite food of Caribou, Musk Ox, and Arctic Rabbit (Fig. 6.4).

FIGURE 6.3 Arctic Moses. *Source: Photo Courtesy A.A. Mohamed Hatha, personal communication.*

FIGURE 6.4 Arctic Willow. *Source: Photo Courtesy A.A. Mohamed Hatha, personal communication.*

1.4 Flowering plants

At the beginning of summers, the ice of the Arctic land starts melting. Although at that time the weather remains very cold (approximately 0°C) with continuous snowstorms, the sun starts shining too. During that time, the flower plants with various colors start growing and blooming. The period from early June to late July is the most favorable for the growth of these plants. The plants include Yellow saxifrage (Fig. 6.5), Purple saxifrage (Fig. 6.6), Mountain Avens (Fig. 6.7), Wild Crocus (Fig. 6.8), Arctic Poppy (Fig. 6.9), Buttercup (Fig. 6.10), Moss Campion (Fig. 6.11), arctic alpine (Fig. 6.12), and others. Like vegetation plants, these flowering plants of the Arctic region are also very

FIGURE 6.5　Yellow saxifrage. *Source: Photo Courtesy Prashant Singh, personal communication.*

FIGURE 6.6　Purple Saxifrage. *Source: https://commons.wikimedia.org/wiki/File:Purpsaxifrage2.jpg.*

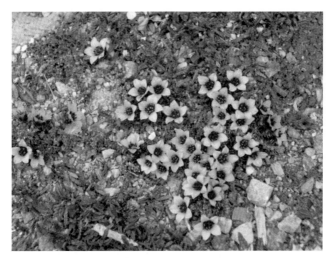

FIGURE 6.7   Mountain Avens. *Source: Photo Courtesy Prashant Singh, personal communication.*

FIGURE 6.8   Wild Crocus. *Source: https://commons.wikimedia.org/wiki/File:%D8%A8%D8%*
B1%D8%AF%D8%A7%D8%B4%D8%AA_%D8%B2%D8%B9%D9%81%D8%B1%D8%A7%
D9%86_%D8%AF%D8%B1_%D8%B1%D9%88%D8%B3%D8%AA%D8%A7%DB%8C_%
D8%B2%D8%A7%D9%84%DB%8C_%D8%B9%DA%A9%D8%B3_%D8%A7%D8%B2_%
D8%A7%D8%AD%D9%85%D8%AF_%D9%86%DB%8C%DA%A9_%DA%AF%D9%81%D8%
AA%D8%A7%D8%B1.JPG.

small. They grow together and remain close to the earth. Stems, leaves, and
buds of certain flowering plants are glabrous, which protect them from the
storm. The seed coat of some plants is wholly adapted to the local conditions.

Flowers of some flowering plants of the Arctic have developed their
morphology as cup shaped to receive a greater amount of solar energy.

FIGURE 6.9 Arctic Poppy- The Arctic poppy is a rare plant species found only in the harsh, Arctic conditions. *Source: Photo Courtesy Prashant Singh, personal communication.*

FIGURE 6.10 Arctic Buttercup. *Source: https://commons.wikimedia.org/wiki/File:Ranunculus_ pedatifidus_IMG_4188_fliksoleie_endalen.JPG.*

Certainly, they face toward the sun when they bloom so that the rays of the sun can focus on their central region. In this way, these plants are warmer than their adjacent air.

Flowers of some plants of the arctic region have developed dark colored pigments, which receive comparatively more solar energy. Though the

FIGURE 6.11 Moss Campion. *Source: Photo Courtesy A.A. Mohamed Hatha, personal communication.*

FIGURE 6.12 Arctic alpine. *Source: https://commons.wikimedia.org/wiki/File:Oxyria_digyna_ IMG_3642_fjellsyre_longyeardalen.JPG.*

sunshine of the Arctic does not have much heat/warmth in it, during summers the flowers of these plants continuously bloom for several months. As there is continuous sunshine in the Arctic region, meaning thereby that there is no complete darkness during the night; hence, these flowering plants develop flowers sooner.

FIGURE 6.13 Cotton grass. *Source: https://upload.wikimedia.org/wikipedia/commons/2/2d/Eriophorum_Cotton_Grass.JPG.*

1.5 Cotton grass

It is a white cottony, ball-like flowered plant, which is the favorite food of the Caribou calf. The calves feed on it and grow very rapidly and become healthy. This cotton grass (Fig. 6.13) plant is also the main food of migratory bird snow-geese.

The Inuit community uses cotton-grass at a large scale. They use the seed apex as a cotton roll in oil lamps. Besides this, they fill these plants in the panties of children, and when the cushion of this bunch gets soiled, they throw it away.

Reference

[1]   http://ete.cet.edu/modules/msese/earthsysflr/tundraP.html.

Chapter 7

Fauna of the Arctic

A large part of the Arctic, including both land and sea, are covered with snow during a maximum period of the year. The animals whether they are terrestrial or aquatic or amphibians, all have acclimatized according to the environment of the Arctic. They are different from the animals in other regions. Nature has provided them the capabilities to survive in extreme climate conditions and live in darkness for months together.

The fauna in the Arctic is divided into three types:

 (i) Herbivorous mammals like Lemmings, Voles, Caribou, Arctic Hares, and Squirrels, etc.
 (ii) Carnivorous mammals like Arctic Foxes, Wolves, and Polar Bears, etc.
(iii) Migratory birds like Ravens, Snow Buntings, Falcons, Loons, Sandpipers, Terns, Snow Birds, and various species of Gulls.

The Arctic is the region settled north of the polar circle. It encompasses North America's Norths, the Greenland region and Asia, and the Arctic Ocean. This ocean is roofed with an Ice pack that never fully thaws. Short plain vegetation grows on the land, wherever the temperature seldom rises higher than 10°C. Despite the acute living conditions, some endemic people like the Eskimo and the Lapps have created their indigenous relations with the Arctic. Solely the animals that have best customized to the cold, just like the bear, white fox, and ringed seal, sleep in these hostile regions year-round. The walrus, seals, and various bird species migrate once the cold intensifies. The Arctic and its inhabitants are sensitive to global warming as well as marine and atmospheric pollution that accumulates within the region, carried there by the winds and ocean currents [1].

Apart from the polar bear, there is a wide variety of animal life, both aquatic and terrestrial. In the Arctic, there are about 130 species of mammals, 280 species of birds, 3000 species of insects, 450 species of fish and reptiles and amphibians. Marine life is also comprised of algae, krill, zooplankton, and microorganisms that are essential for the survival of all other species, even as large as whales. It is amazing to know that a large variety of organisms live there. And although it seems a lot, it is said that there are still more sea creatures that have not yet been discovered in the Arctic Ocean, since weather conditions make exploration very difficult and only limited parts have been studied.

The Arctic. https://doi.org/10.1016/B978-0-12-823735-9.00007-2

There are abundant species of animals in land and sea regions of the Arctic. The animals can be divided into terrestrial, aquatic, and amphibians. The number of migratory birds that come during summers to the Artic and permanent dwellers of the Arctic is also very large. Among the terrestrial animals of the Arctic, the names that appear spontaneously in mind are Caribou and Polar Bear. Polar Bear due to its white color it is also called as white bear. Other Arctic animals include Fox, Rabbit, Musk-Ox, Laming, Stoat, and Wolverine.

1. Caribou

Caribou is originally a wild animal, but it has been domesticated. Its domesticated species is called "Reindeer." It is said that approximately 10 lakh years ago in the Pleistocene era, large groups of Caribou were found in Northeast America and Europe. The pictures of Caribou (Fig. 7.1) found among the different animals carved in the caves of France (25−30 1000 years ago) lead to the speculations that during this time Caribou was domesticated. Although their eye-sight is weak, they have a strong smelling sense with developed hearing capacity.

Caribou is called as *Rangifer Caribou* by zoologists. It belongs to the clan of deer that lives in large flocks. It is stout and strong with a magnificent appearance. It can bear a load of 90 kg and can travel 65 km in a day. It can walk easily on ice or wetland. Even it can pull a load of 200 kg with a speed of 30 km/h when yoked in a sledge.

FIGURE 7.1 Caribou. *Photo Courtesy Mohammad A. Hatha.*

2. Polar Bear

If we want to exhibit the Arctic as a symbol, the symbol will be Polar Bear. The picture or image of Polar Bear (Fig. 7.2) reminds us of the Arctic. Some people believe that the Polar Bear (*Ursus Maritimus*) is a terrestrial mammalian who spends most of its time floating on ice pieces of the Arctic Ocean. On the contrary, some believe that it is an aquatic animal that lives on land for maximum time. It wanders on floating ice pieces in search of its favorite food, that is, ringed-seal. Though, it can also be seen wandering in thickly populated areas searching for food in the heaps of leftover foodstuff. Thus, it is a wandering animal that is present both on land and water in search of food. Polar Bear is larger than other species of Bears. It looks very magnificent, beautiful, and attractive. It is very strong and powerful and most of the animals it preys on; dies in a single attack by its strong claw. Males are larger than females in size. The weight of an adult male is 400−600 kg. Although Polar Bear is mainly a Carnivorous, it can feed on plants and berries occasionally. Sometimes in search of food, it reaches to the kitchens of the local people and at times it snatches the food being cooked by them. Its smelling power is very strong. So, people who live near the habitat of Polar Bear are asked to cover the food properly.

It does not have any inhibition to eat the dead animals and in the end of summers or during autumn it can be seen eating the dead whales and walruses lying at the seashore. Polar Bear is not a cannibal by nature rather afraid of human beings, but if instigated it can attack, which can be fatal to humans. Ringed Seal (*Pusa hispida*) is the favorite food of Polar Bear and it keeps wandering and floating on the ice pieces in search of it. The color of its fur

FIGURE 7.2 Polar Bear. *Photo Courtesy Mohammad A. Hatha.*

matches well with the ice. That is why the seal lying on the ice mass gets deceived by the quietly sitting Polar Bear and falls prey to it. Due to its strong-smelling power, the Polar Bear can observe the seal hidden under ice and tries to find out the hole by which the seal breaths. Ringed Seal is an aquatic animal, but it is a mammal and it cannot utilize the dissolved oxygen of water. It requires free O_2 present in the air for its respiration. For this, it requires such holes in the ice to get the oxygen from the atmosphere.

When the Polar Bear finds such holes, it sits there if the seal extrudes its head out. It avails the opportunity and pulls out the seal to eat. It is strange that it removes its skin, fat, and internal organs of Ringed Seal but does not eat the meat. During summers, due to the melting of ice, the seal goes under the water and it becomes difficult for the Polar Bear to catch it. So, it feeds himself sufficiently before the summer season. In fact, to conserve its energy, Polar Bear must eat 50−75 seals in a year. But it is surprising that on some occasions, it can survive up to 6 months without any food. Although Polar bear is quite a swift animal, it can run fast and shortly attains a speed of 30 km/h, but it gets tired very soon. After being chased by Snow Mobile, its body heat increases, and it can die of heatstroke. Polar Bear is a good swimmer and can swim for long in the cold waters of the Arctic. There is no danger of freezing of its organs because of the oily skin with fur and the thick layer of fat below the skin that saves it. After coming out of the water, it sweeps off the water droplets by shaking off its fur. The feet of Polar Bear is like snowshoes. The hair grown on its underfoot help the bear to walk easily on the slippery snow. To protect itself from slipping, it retracts its paws inside.

According to scientists Polar Bears are being affected by global warming and melting of ice of the Arctic region. Therefore, scientists are finding ways and means to protect them. The Inuit community believes that Polar Bear is intelligent, strong, and like human beings. There are a few local tales that Polar Bears were like human beings who used to live in Igloos and communicate with each other.

3. Musk Ox

The literal meaning of Musk Ox (Fig. 7.3) is "Kasturi Bull" but unlike our "Musk Deer," Musk Ox, this habitat of Arctic does not have Musk in it. According to scientists, it is a mixed breed of Cattle and Bison. Its zoological name is *Ovesmuscatus* (means Sheep Bull). It has brown long dense hair hanging to the ground. A layer of soft hairy lining lies below the layer of hair. The hair and this layer protect the "Musk Ox" from the terrible cold of the Arctic. Its hair falls during summers. The Inuit people make wool from these hairs. Both male and female Musk Ox has horns. They live in a group of 10−20 and create a unique sound "burburburbur" when irritated. They make a defensive boundary when there is doubt of attack by enemies (Carnivores). The adults are outside, and the calves are inside the boundary. To protect

FIGURE 7.3   Musk Ox. *Source: https://upload.wikimedia.org/wikipedia/commons/e/ed/Ovibos_moschatus_qtl3.jpg.*

themselves and their calves, sometimes the Musk Ox attacks their enemies by hitting with their horns. The dangerous wolves of the Arctic are their main enemies. It gets tired if it must run for long due to increased body heat.

The main food of Musk Ox is Grass, Lichen, and Willow. The flocks keep wandering in search of food. They do not graze by standing in one place, they keep moving while grazing.

4. Fox

Arctic Fox (Fig. 7.4) is mainly found near the region of Arctic Circle in the farthest north. In other words, the Northern part of Alaska, Nunavut Island, and Coastal areas of Greenland are their habitats. It is very difficult to identify the fox in winters because its thick white fur makes it a camouflage in ice.

Although it is equal to the normal cat in height, its thick fur makes it higher in size. Interesting thing is that the color of the fur is white in winters, but it becomes turning to gray with the decreasing cold and in summers it turns brownish gray.

The dens of fox families are close to each other and in winter they dig underground passages to get connected with each other. Usually, during winter, they also like hunting together. They prey Lamingo mostly. They also eat Rabbits, Squirrel, Birds, and their eggs. If the Arctic fox finds the baby seals, it kills and eats them. The Arctic fox is also cunning like the other foxes of the world. While walking on ice, it also senses the presence of prey below the ice. If it gets even a small hint of prey, it breaks the ice by jumping over it and catches the prey. During summers, it hides its food in the den or below any stone/rock. It is a unique way to protect its food.

FIGURE 7.4 Arctic Fox. *Photo Courtesy: Mohammad A. Hatha.*

5. The Arctic Hare

The Arctic Hare (Fig. 7.5) (*Lepus arcticus*) lives in Greenland and North America's far north. It digs its shelter in the snow to sleep and to protect itself from the wind and intense cold of arctic winter. Its hind feet are lined in dense fur to assist preserve body heat. When it senses a predator like a wolf or fox approaching, this small mammal rises up on its hind legs and can bound away at a speed of up to 50 km/h.

FIGURE 7.5 Arctic Hare. *Source: https://commons.wikimedia.org/wiki/File:Arctic_Hare_1.jpg.*

6. Snowshoe Hare

Snowshoe Hare (*Lepus americanus*) are found in Canada, Greenland, and Islands of the Arctic Ocean. They live in the hilly region as well as plains. They are herbivorous animals, so they like to live in regions where there are plants with faster growth. In winters, the Rabbits go to such places where they do not have to dig the ice for food. The rabbit has a thick fur coat with thick gray-colored layer of skin below it. The fur coat protects it from cold. During the winters, the color of the fur becomes white (Figs. 7.6 and 7.7).

7. Wolverine

Wolverine (*GuloGulo*) (Fig. 7.8) is an animal of the Bezel family that frightens other hunting animals. It is found in every part of Canada including parts of the Arctic region. It is equal to the size of a baby Bear but is a stronger and more dangerous animal. Its legs are very strong, and its paws are very big. Though it is very strong, it rarely hunts by itself rather lets other animals to hunt. When they kill the animal, it reaches there growling and showing its ugly teeth to chase away the hunter animals. Wolverine eats the prey itself later.

Wolverine's main food includes rats, small mammals, birds and eggs.

8. Weasel (*Mustela nivalis*)

Weasel (Fig. 7.9) is a small animal measuring 15−20 cm in height having a small face with sharp nails and teeth. The color of its fur coat is white that turns brown in summer but the hair at the end of its tail always remains black. Although Weasel is a small animal, it is always ready to fight with anyone and

FIGURE 7.6   Snowshoe Hare. *Source: https://commons.wikimedia.org/wiki/File:Lepus_americanus_5459_cropped.jpg.*

FIGURE 7.7 Another color of Snowshoe Hare. *Source: https://upload.wikimedia.org/wikipedia/ commons/3/38/Snowshoe_Hare%2C_Shirleys_Bay.jpg.*

FIGURE 7.8 Wolverine. *Source: https://upload.wikimedia.org/wikipedia/commons/0/0a/Gulo_ gulo_2.jpg.*

swoop at once on its enemy who might be much larger than its size. The babies of Weasel are very clever in this way. They start preying when they are 2 months old only. Weasel eats rabbits, rats, birds, frogs, squirrels, and pica (a small rodent of the Arctic region).

FIGURE 7.9   Weasel. *Source: https://upload.wikimedia.org/wikipedia/commons/e/e3/Mustela_ nivalis_-British_Wildlife_Centre-4.jpg.*

Ermine is the smallest Weasel. It is also called as Stoat. Ermine likes to live anywhere it can make a burrow, can store food, and can feed babies. Generally, the burrows of Ermine are heaps of unbound soil or stones. Its main food is lemmings so when the population of lemmings increases, simultaneously the population of Ermine also increases.

9. Lemming

There are two types of Lemmings in the Arctic, Brown, and Color. The Brown Lemming lives in moist areas and the color Lemming in hilly regions. The fur of color Lemming turns white in winters. Lemming (Fig. 7.10) lives in treeless

FIGURE 7.10   Lemming. *Source: https://upload.wikimedia.org/wikipedia/commons/f/f2/Dicrostonyx_ torquatus.jpg.*

regions. They eat grass. In winters, they make a burrow in ice and live there. For this, they make many tunnels and by entering into them, they protect themselves from high winds and cold. Generally, they also eat roots of plants, fruits, and Lichens, etc.

10. Wolf

Mainly two types of Wolves are found in the Arctic region, Tundra Wolf (Fig. 7.11) and Arctic Wolf (Fig. 7.12). Tundra Wolves are found in the mainland region. Generally, they are brown or gray colored and prey on Caribou. On the contrary, Arctic wolves live in the Northern Islands. They are comparatively small and white-colored and mainly prey on Musk Ox and Arctic Rabbit. Both type of wolves' lives in family and besides their main prey, they also eat fishes, lemmings, foxes, squirrel, etc.

11. Arctic Squirrel

The Arctic (Fig. 7.13) Squirrels live under the earth making burrows in which there are many tunnels and rooms. They live in groups and eat plants, seeds, and fruits in large amount during summers. They also store the food for difficult times. Due to excessive eating, the extrafat deposited in their body is utilized during winters when they are in the hibernation period. As far as the question of stored food is concerned, it is proved useful when they awake from hibernation during spring. They eat them till the new plants grow in the Arctic region.

FIGURE 7.11   Tundra Wolf. *Source: https://upload.wikimedia.org/wikipedia/commons/c/c9/%D0%92%D0%BE%D0%BB%D0%BA_3.jpg.*

FIGURE 7.12 Arctic Wolf. *Source: https://commons.wikimedia.org/wiki/File:Polarwolf004.jpg.*

FIGURE 7.13 Arctic Squirrel. *Source: https://upload.wikimedia.org/wikipedia/commons/4/4a/ Spermophilus_parryii_%28eating_mushroom%29.jpg.*

It will be important to mention here that Arctic Squirrel is the only animal that goes for hibernation in winters. For this, the squirrel makes a layer of Lichens, leaf, and hair of Musk Ox in its burrow. Rolling on it like a ball, wrap it on its body and go for hibernation for several months during winters. During this period of hibernation, the temperature of its body falls up to freezing level and the heartbeat becomes very slow.

12. Birds in Arctic

Birds return to the Arctic region when the spring season starts. Birds usually migrate to warmer regions to protect themselves from cold. The return of these birds is as pleasant as the return of the sun. During winters, the Eskimo must confine to the meat of animals and fishes but now they can taste the eggs of snow-geese and other birds.

The birds who return with the spring are Arctic Tern, Golden Plover, Wild Ducks, Geese, Pintail, Mallard, and other species of birds. There are some birds that are a permanent inhabitant of the Arctic like Snow Owl, Raven, Ptarmigan, and Gyrfalcon.

13. Arctic Tern

Among the birds of the Arctic region, perhaps Arctic Tern (*Sterena paradiscii* or *Sterena macroos*) is entirely peculiar. The migratory journey of this small beautiful bird is the longest not only in birds but in the whole animal kingdom. It wants to enjoy the summers of the Arctic region as well as that of the Antarctic region. In this way, its migratory journey starts from the Antarctic to the Atlantic and back which is 32,000 km. While doing this, the Arctic Tern remains in sunlight for 24 h during 8 months of the year. Although it breeds in Arctic region only. The color of Arctic Tern (Fig. 7.14) is white, its back is gray, and the upper part of the head is black. The color of the beak and feet is red. Its feathers are long, and the tail is divided into two parts.

It makes the flight of thousands of kilometers over the seas. That is why it catches its food during flying (flying insects.). Generally, it dives straight into the sea to catch a fish for food. Female lays two to three eggs at a time. Arctic Tern's babies remain hidden in such a way in the nest that they cannot be seen from outside. The life span of Arctic Tern is quite long, and they live up to 34 years.

FIGURE 7.14 Arctic Tern. *Photo Courtesy Shabnam Choudhary.*

14. The Snow Goose

In summer, the Snow Goose (Fig. 7.15) nests in large colonies on the Arctic tundra, eating grass, berries, and seeds. When fall comes, it migrates to the south in giant flocks to Mexico and the Southern United States, where it spends winter living on the plains and in the fields. The snow goose resumes its journey and heads back to the Canadian−North in the spring.

15. Golden Plover

Golden Plover (Fig. 7.16) is the bird whose name comes second in terms of length of the migratory journey, after Arctic Tern. It spends the summers in Aleutian Islands and winters in 3200 km away in Hawaii Islands. It is said that after resting in Choral Islands of Pacific Ocean, some birds reach till New Zealand.

16. Snow Owl

Arctic Owl/Snow Owl (Fig. 7.17) is among those birds that stay in the Arctic despite the extreme cold. The female Snow Owl (*Nicatia Scandiaca*) is larger than their male counterparts. Females are also not full white in color like males. Hairs are found on the bodies of both male and female, which protect them from cold. Snow owl is a carnivorous bird. During flight, it targets its prey (Lemming, squirrel, rabbits, and small birds) and attacks in one go.

Generally, Lemming is its favorite food.

FIGURE 7.15 Snow Goose. *Photo Courtesy Mohammad A. Hatha.*

FIGURE 7.16 Golden Plover. *Source: https://commons.wikimedia.org/wiki/File:Pluvialis_fulva_-Bering_Land_Bridge_National_Preserve,_Alaska,_USA-8.jpg.*

FIGURE 7.17 Snow Owl. *Source: https://upload.wikimedia.org/wikipedia/commons/6/6d/Snowy_Owl_%28240866707%29.jpeg.*

17. Common Raven

Raven (Fig. 7.18) is a bird belonging to the crow family, and it is the largest bird of that family. Like our Indian crow, it is also a clever bird that knocks its beak to check things. It steals the food of other animals and hides away. It eats both vegetarian and nonvegetarian food, and thus is an omnivore. Small mammalian animals, birds, insects, berries, etc. are its favorite food. So, the Inuit people consider it as a "ghost." Its color does not change with the season. It always remains black. It is a quick bird and can fly long. During flight, it can jump and can make turns in the air.

FIGURE 7.18 Common Raven. *Source: https://upload.wikimedia.org/wikipedia/commons/3/30/Corvus_corax.001_-_Tower_of_London.JPG.*

18. Gyrfalcon

Gyrfalcon of Arctic is also a bird of Falconidae family like our Indian Hawk, and it eats small birds like ptarmigan, small mammalians, and squirrel. Its vision is very sharp, and the beak is so sharp that it can tear its own body. Gyrfalcon (Fig. 7.19) is a big bird, and the expansion of its feather may spread up to 1.35 m when fully stretched. Its life span is around 13 years. It also lives in the Arctic region for the whole year.

During middle age times, Gyrfalcon was known as Bird of Kings and the kings used to hunt with its help. It is said the all the Gyrfalcons of Iceland were considered as the property of the King of Denmark (Earlier Iceland was under the rule of Denmark).

19. Rock Ptarmigan

Rock Ptarmigan (Fig. 7.20) is a bird of the size of a chicken that also lives in the Arctic during winter. Although it possesses feathers, it flies very less. It bears hair on its legs so that it can run on ice easily. It sheds its feathers many times in a year but every time the color of new feathers is according to the environment of the Arctic. The function of feathers is to protect themselves very well from enemies. It is the belief of Inuit people that the burning of the hair of Ptarmigan causes the rain accompanied by the noise of thunder during winters.

FIGURE 7.19 Gyrfalcon. *Source: https://upload.wikimedia.org/wikipedia/commons/c/c8/Falco_rusticolus_white_cropped.jpg.*

FIGURE 7.20 Rock Ptarmigan. *Source: https://upload.wikimedia.org/wikipedia/commons/1/1f/Rock_Ptarmigan_%28Lagopus_Muta%29.jpg.*

20. Fish

Climate change has threatened marine species of the Arctic Fisheries [2]. The state of currently available scientific knowledge needs to be improved to reduce the substantial uncertainties associated with Arctic fish stocks along with Interim Measures on Control of Commercial Fishing. The range of some sub-Arctic fish stocks is likely to extend into Arctic areas, and fishing activity will increase in future due to decreasing ice-cover [3].

FIGURE 7.21 Krill. *Source: https://upload.wikimedia.org/wikipedia/commons/e/e3/Meganycti phanes_norvegica2.jpg.*

21. Krill

The phytoplanktons grow its population simply below favorable conditions; however, they are consumed by shrimp like malacostracan crustacean. In the Norwegian language meaning of malacostracan crustacean is "little fish" however malacostracan crustacean is not a fish.

Krill is concerning 2 in. long associated with an antenna like an organ of 2–3 cm on its body. Krill (Fig. 7.21) feeds on flora, microscopic, noncellular plants that drift close to the ocean's surface and live off dioxide and sun's rays. Its life is around 6–7 years. There is a stunning truth about *malacostracan* crustacean that it will survive without food for a year. These tiny, shrimp-like crustaceans' square measure primarily the fuel that runs the engine of the Earth's marine ecosystems.

22. Whale

Whale will sustain and board the cold water of Arctic terribly simply. Usually, their square measure three kinds of whale—Beluga (Fig. 7.22, whalebone whale (Fig. 7.23), and Narwhale (Fig. 7.24).

In Russian Language, Beluga suggests "White Animal." The color of the adult whale is white. The form of its face is specified as it appears smiling. Beluga makes voices like shrieks, whistles, or chirps like birds. That's why it is conjointly known as "Sea Cannery." It comes back on the surface of the ocean to breathe by breaking the floating ice packs with its head. Narwhale is additionally known as "Sea Unicorn" (Fig. 7.25).

FIGURE 7.22 Beluga whale. *Source: https://upload.wikimedia.org/wikipedia/commons/1/19/Beluga03.jpg.*

FIGURE 7.23 Whalebone whale. *Source: https://upload.wikimedia.org/wikipedia/commons/9/9e/Humpback_stellwagen_edit.jpg.*

FIGURE 7.24 Narwhale. *Source: https://upload.wikimedia.org/wikipedia/commons/e/e6/Narwhalsk.jpg.*

FIGURE 7.25 Sea Unicorn (Narwhal) in Arctic. *Source: https://upload.wikimedia.org/wikipedia/commons/e/e6/Narwhalsk.jpg.*

23. Walrus

It is a vertebrate that has a shapeless body with long and nailed grins and it is troublesome to maneuver toward land. On the contrary, swims swiftly within the ocean and might dive to its bottom level owing to its streamlined body. Its grins facilitate in initiating off the ocean and jump on the rocks.

It has thick skin with blubber layers on its body to protect it from the extreme climate of the Arctic. Therefore, it can easily lie on the ice. It has wrinkled brown skin. Because of the heat of the sun, its blood comes near its skin and it becomes red.

Walrus (Fig. 7.26) eats Clam, Krill, Crabs, and other small creatures. It also feeds on fish and octopus. It detects is prey in shallow water. Its two special water bags help him in swimming. One is under the throat and another under the neck. Walrus fills the water bag structures with air and then use them as floaters.

FIGURE 7.26 Walrus. *Source: https://upload.wikimedia.org/wikipedia/commons/8/8f/Walrus_2_%286383855895%29.jpg.*

24. Seal

Seal is a mammal. Arctic (Fig. 7.27) has three types of Seal—Harp Seal, Hooded Seal, and Ringed Seal. Because of the ringed shapes on its skin, the seal is called Ringed seal. It is the smallest seal with one-and-a-half-meter length. It is the favorite food of Polar Bear. The Eskimos also depends on it for their food.

The female Ringed Seal comes on land or on ice pack for breeding. It chews the ice on the layer of the sea to make a hole for breathing. The Polar Bear and Arctic Fox are its enemies.

25. Marine living resources

With reference to the event of the development of unexploited marine living resources, it is necessary to push through due cooperation with the coastal states and to secure the necessity for food security in a balanced manner whereas making certain the property of the resources supported scientific evidence. The Law of the Sea, framed by the United Nations Convention has the right to use Arctic's natural resources in a beneficial way among their exclusive economic zones: Russia, Canada, Denmark, Norway, and the United States (though the United States has however to approve the treaty as it considers the agreement to be customary law of nations and abides by it). The Arctic region and its resources have recently been at the center of controversy and pose potential conflicts between nations that have differing opinions of how to manage the area, including conflicting territorial claims. In addition, the Arctic region is home to an estimated 400,000 indigenous people.

FIGURE 7.27 Arctic Seal. *Photo Courtesy: Shabnam Choudhary.*

These indigenous places would be at risk and need to be displaced if the ice continues to melt at the current rate. This will contribute to climate change accelerates ice depletion that releases methane. Ice reflects incoming solar radiation, and without it, it will cause the ocean to absorb more radiation (albedo effect), heating up the water causing more ocean acidification, and melting ice will cause a rise in sea level.

References

[1] http://www.ikonet.com/en/visualdictionary/static/us/fauna_flora_arctic_region.
[2] https://www.newsdeeply.com/oceans/articles/2017/05/12/ocean-warming-is-already-affecting-arctic-fish-and-birds-2.
[3] https://en.wikipedia.org/wiki/Arctic_policy_of_the_United_States#cite_note-Bush-60.

Chapter 8

Nonliving natural resources

The nonliving natural resources of the Arctic are the hydrocarbons (oil and gas), minerals resources, and Gas Hydrate deposits within the Arctic Circle that can be commercially benefitted to humans. The mineral resources include major reserves of oil and natural gas, large quantities of minerals including iron ore, copper, nickel, zinc phosphates, and diamonds apart from abundant Gas Hydrate potential of the Arctic Ocean.

Ever since the commercial extraction of oil that began in the 1920s in Northwestern Canada and subsequent impetus to oil exploration in the Arctic region during the 1960s, numerous hydrocarbon fields were discovered in Yamalo-Nenets region (Russia), the North Slope of the Brooks Range (Alaska), and Mackenzie Delta (Canada). Billions of cubic meters of oil and gas have been exploited by the countries having territorial claims over the Arctic region such as Russia, Alaska, Norway, and Canada.

It is noteworthy that about 400 odd onshore oil and natural gas fields are situated north of the Arctic Circle and more than two-thirds of the producing oil and gas fields are discovered in western Siberia (Russia). Khanty-Mansiysk Autonomous Region of Russia is the largest producer of petroleum oil (about 57% of total oil production of Russia). The world's biggest Shtokman oil field is reported to have about 4000 billion cubic meters of natural gas.

According to an estimate by the Arctic US, the oil reserves are about 15 million barrels; whereas, natural gas reserves are 2 trillion cubic meters. About 20% of total oil production is only from Prudhoe Bay Oil Field. Similarly, Arctic Canada has 49 oil and gas fields in Mackenzie River Delta while 15 such fields are located over the Arctic Archipelago of Canada. Nevertheless, the biggest gas fields are confined to coastal Alaska and Siberia.

The Arctic holds an important portion of the world's undiscovered petroleum resources, both in terms of hydrocarbons and minerals. It has historically been a difficult endeavor to exploit these resources due to the natural barriers created by harsh weather conditions and difficult terrain [1]. In many parts of the Arctic, scientists have got much success in the exploration of minerals, petroleum, natural gas, and gas hydrate reserves.

The resources are also unevenly distributed, for instance, the Russian region is richer in gas reserves, while the Norwegian region is rich in oil resources [2].

The Arctic. https://doi.org/10.1016/B978-0-12-823735-9.00011-4

Below the earth surface somewhere up to 100 m deep, due to the presence of snow layers and very harsh climate and difficulties of commuting, etc. the exploration of minerals in the Arctic region could never be done as satisfactorily and in a well-organized manner as in other areas of the world. The Russians have made many attempts in the Siberian region and they have also been successful.

Exploration, development of production, and transport facilities for oil, gas, and mineral resources are increasing throughout the circumpolar region. Receding sea ice cover and permafrost thaw will influence accessibility to mineral and energy resources both on land and in the Continental Shelf in the future. Owing to the huge potential oil and gas resources as well as coal, minerals, and diamonds, the regions around the Arctic basin are receiving the attention of all bordering countries.

1. Hydrocarbons potentials of Arctic

The Arctic is known to have wealth of petroleum and other natural resources. This region is reported to have produced about tenth of the world's oil and one-fourth of its natural gas. It is significant to know that about 80% of oil along with all resources of natural gas is found in the Russian Arctic. Whereas, other leading producers of natural oil are Arctic Canada, Alaska, and Norway. Fig. 8.1 highlights the spatial potential of oil and gas in the Arctic region; whereas, Fig. 8.2 shows the potential of Russia, Canada, Greenland (through Alaska), Norway, and the United States about their oil and gas resources potential in their respective Arctic region.

2. Oil and gas

Petroleum and natural gases are not only found in land but also in the bottom of the Arctic Ocean and they are being extracted on large scale for industrial use. Below the land and coastal areas of Alaska, Canada, Greenland, Scandinavian countries, and Siberia, large reserves of petroleum and natural gases are present. Alaska is famous in the whole world for its petroleum reserves and residents of Greenland are happy that it will be possible to extract petroleum from large stores when due to the result of global warming the land will become free from ice.

Arctic region is composed of 19 geological basins such as North Slope, Beaufort Sea, South Arctic Islands, Franklinian Sendrup, Baffin Bay, Labrador Shelf, Southwest Greenland, North Greenland, Kronprins Christian Basin, West Barents Sea, East Barents Sea, North Kara Sea, South Kara Sea, Laptev Sea, East Siberian Sea, Hope Basin, North Chukchi Sea, and Pechora Sea. Some of these basins have experienced oil and gas exploration for example the Alaska North Slope first produced oil in 1968 from Prudhoe Bay. However, only half the basins such as the Beaufort Sea and the West Barents Sea have

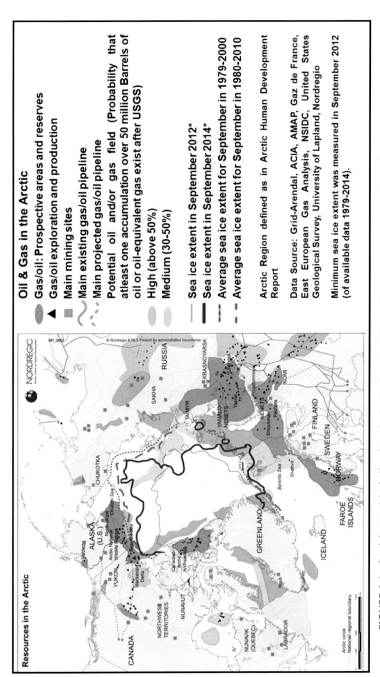

FIGURE 8.1   Spatial potential of oil and gas in the Arctic region. *Source: https://www.nordregio.org/maps/resources-in-the-arctic.*

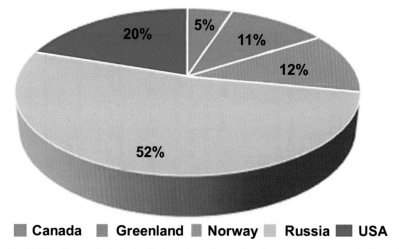

Potential Arctic Oil and Gas Resources
(Total Assessed Resources = 412 Billion boe)

FIGURE 8.2 Potential of Arctic Nations about their oil and gas resources potential in their respective Arctic region.

been explored and thus carrying lots of unexplored hydrocarbon reserves in the Arctic region. Fig. 8.3A–B shows location of various basins in the Arctic region as assessed by the United States Geological Survey (USGS).

According to a Geo-scientific survey, Department of the United States; in the bottom of the Arctic Ocean more than 1800-million-barrel petroleum is situated, between Greenland and Canada, and 3100 million barrels is situated in the eastern coast of Greenland itself. The shelf area of Norway's continent is very large, and its area is 30% of the whole continental shelf of Europe. The extraction work of petroleum had been started at the beginning of a decade of 1970, and today Norway is the third-largest exporter of crude oil in the world and it is also supplying natural gas to most of the European countries. Located in the North of Norway and falling under the exclusive economic zone, Barents Sea has a storage capacity of 85 crores MTOE (Million tonnes oil equivalent—10 lakh tons oil) of petroleum and natural gas. If we consider the storage capacity of petroleum and natural gas at the bottom of the Barents Sea situated in the exclusive economic zone of Russia, it is more in comparison to Norway's 450 crores MTOE reserves. Large reserves of petroleum and natural gas are also found near the Northern coast of Siberia.

In the late 2000s, receding sea ice made most of the regions accessible due to which interest in offshore hydrocarbons had increased. As the global energy demand increased, US Government estimated huge probable oil and gas

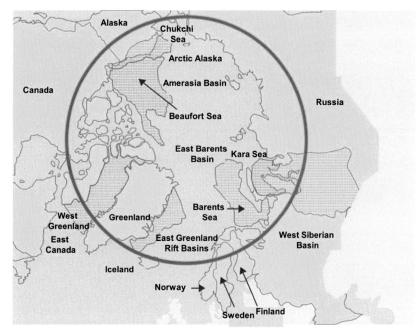

FIGURE 8.3A Resource basin in the Arctic circle. *Source: https://upload.wikimedia.org/ wikipedia/commons/7/7d/The_Resource_Basin_in_the_Arctic_Circle_.jpg.*

reserves throughout the Arctic, and an additional politically stable investment climate relative to different international regions with giant organic compound resources. These factors have spurred the Arctic coastal states to support offshore oil and gas development.

Almost 22% of the world's oil and natural gas could be located beneath the Arctic estimated by the USGS in its studies. The estimate of oil availability to the United States is in the range of 30 billion barrels while that of natural gas reserves could be about 221 billion cubic feet. Historically, challenging arctic conditions presented a great risk of developing extraction infrastructure. However, with global warming affecting the arctic disproportionately to the rest of the world, and with sea ice covers and flows subsiding by as much as 30% over the last 30 years, these resources are now becoming more accessible. Beyond oil and gas, however, the Arctic coast is also home to the Northern Sea Route, which Russia hopes will 1 day be used to ship hydrocarbons and other goods to and between Asia and Europe. The country also foresees that increased sea access will help the development of Russian regions further inland [3].

Beginning with resources, the Arctic is home to an estimated 412 billion barrels of recoverable and conventional oil, natural gas, and natural gas liquids. Approximately one-third of the territory consists of continental shelves

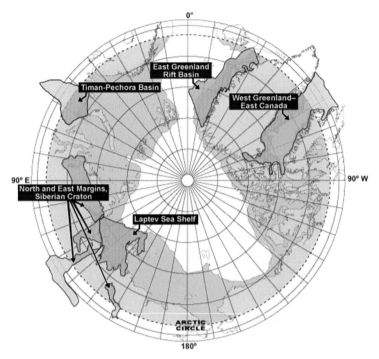

FIGURE 8.3B Location of various petroleum. *Source: https://upload.wikimedia.org/wikipedia/commons/1/1f/ArcticLocationMap2.png.*

that have vast areas to be explored. 87% of these resources are thought to lie in seven Arctic basin provinces, as shown in Table 8.1.

Currently, the region produces concerning one-tenth of the world's oil and one-fourth of its fossil fuel. The Russian Arctic supply 80% of this oil and nearly all the fossil fuel. Arctic North American nation, American state, and Norway are the opposite leading producers [4]. Recent appraisals suggest that a considerable fraction of the world's undiscovered petroleum reserves lie within the Arctic.

In recent years, both Arctic and non-Arctic nations have shown a growing interest in the potential of the Arctic for the extraction of hydrocarbons. Of the Arctic states, Russia has invested most heavily in oil and gas exploration in the region not surprising, given that 43 of the 61 large oil and natural-gas fields already discovered in the Arctic are located within its exclusive economic zone (EEZ). This compares to only 11 in Canada, six in the United States, one in Norway, and none in Greenland (Denmark). In addition to regulatory issues, the lack of existing infrastructure has been an obstacle to oil and gas companies looking to invest in the Arctic. The harsh climate creates a difficult operating environment, and companies need equipment specially designed for

TABLE 8.1 Outlining the total crude oil, natural gas, and natural gas liquids oil equivalent in billions of barrels [3].

Petroleum province	Crude oil (billion barrels)	Natural gas (trillion cubic feet)	Natural gas liquids (billion barrels)	Total (oil equivalent in billions of barrels)
Wet Siberian basin	3.66	651.50	20.33	132.57
Arctic Alaska	29.96	221.40	5.90	72.77
East Barents basin	7.41	317.56	1.42	61.76
East Greenland Rift basin	8.90	86.18	8.12	31.39
Yenisey-Khatanga basin	5.58	99.96	2.68	24.92
Amerasian basin	9.72	56.89	0.54	19.75
West Greeland-East Canada	7.27	51.82	1.15	17.06

the unpredictable yearly polar ice floes and deep-sea beds. Current offshore drilling structures can only reach hydrocarbons that lie in deposits less than 100 m deep, but one-third of the Arctic comprisesof ocean waters projected to be more than 500 m deep. Many countries also do not have the necessary infrastructure, such as required drilling platforms, power plants, storage tanks, pipelines, and transportation, to extract and transport the oil and natural gas.

Lack of necessary infrastructure not only affects the economic viability of such projects, but also the environmental liabilities that a company would take on by drilling in the Arctic. Given these challenges, oil companies may be deterred from investing in the Arctic, which could hinder the development necessary to exploit hydrocarbons in the medium term. There is also the danger of these extraction activities triggering negative consequences such as oil spills.

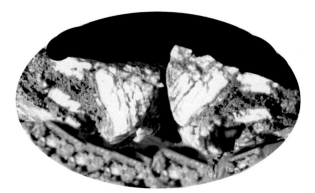

FIGURE 8.4 Gas-hydrates recovered from Lake Baikal Indo-Russian long-term program. *Photo Courtesy Kala Chand Sain.*

3. Arctic natural gas hydrates

These gas hydrates are medium to high-pressure stabilized, ice-like compounds stabilized by higher than sea level ambient pressure found in the cold, deep-ocean environment. The specific structure of a gas hydrate piece, from the subduction zone off Oregon is shown in Fig. 8.4. A typical clathrate structure is depicted in Fig. 8.5.

Natural gas, primarily methane, is held within a water molecule crystal lattice and thermodynamically stabilizes the structure through hydrogen bonding. Gas is concentrated by entering a crystalline structure: from 140 to

FIGURE 8.5 A typical clathrate structure (gas hydrate). *Source: https://commons.wikimedia.org/wiki/File:Burning_hydrate_inlay_US_Office_Naval_Research.jpg.*

200 volumes of gas can be concentrated in one volume of the hydrate. Hydrate is thermodynamically stable in the Hydrate Stability Zone Hydrate formation, and trapping of gas as a long-term part of marine basin genesis is a first-order concentration process that can trap immense volumes of gas in marine sediment.

Hydrates probably occur over broad areas of the Arctic Ocean and northern Nordic Sea. By analogy with identifications elsewhere, bathymetrically controlled shallow gas reservoirs also can be expected within 1.5 km of the sediment surface within the thicker Arctic Ocean sediments. Hydrate and trapped gas in thick sedimentary areas in the Arctic and Nordic seas have been identified locally in the western Barents Sea and northern Alaska continental margins. Analysis of sediment thickness, sediment types and sedimentation history, and heat flow suggests that widespread generation of gas has been trapped in, and possibly below impermeable hydrate, over 1.5×10^{6} km^2 of the Arctic and Atlantic Polar region. Hydrate can be expected to occur in sediments of most of the Canada and Wrangel basins, and in most of the continental slope and rise sediments. These hydrates comprise one of the planet's major energy resources.

4. Mineral resources

The Arctic holds a rich wealth of minerals, including phosphate, bauxite, iron ore, copper, and nickel being abundantly used in various industries. About 11 metric tons of phosphates, which is 8% of the global output is produced by Russia alone. The Arctic mineral resources include apart from major reserves of oil and natural gas, large quantities of minerals including iron ore, copper, nickel, zinc phosphates, and diamonds.

The Russian Arctic, being the most developed sector of the region, holds abundant deposits of nickel, copper, coal, gold, uranium, tungsten, and diamonds. On the contrary, the North American Arctic contains pockets of uranium, copper, nickel, iron, natural gas, and oil. With global warming, a huge quantity of diamonds, gold, nickel, and other metals in the Arctic are expected to be exploited in the Arctic region. However, at present, only six mines are operational in North America's vast arctic. The Arctic region consists of large quantities of minerals, including phosphate, bauxite, iron ore, copper, and nickel. These minerals have great use in industrialized economies. Russia produces on an average of 11 M tons of phosphates [5].

There are several large mines in the Arctic region that include Red Dog mine (zinc) in Alaska, Diavik Diamond Mine in Northwest Territories, Canada, and Sveagruva in Svalbard. Also, quite a few large mines under development are Baffinland Iron Mine in Nunavut and Isua Iron Mine in Greenland. In Alaska, Gold mines are widespread. Fort Knox Gold Mine is the largest producer of gold in the history of Alaska.

The Arctic is also rich in minerals, such as nickel and copper ore. Mineral resources also include gemstones and rare earth elements, which are used in batteries, magnets, and scanners. Gold is found and has been mined throughout Alaska; except in the vast swamps of the Yukon Flats, and along the North Slope between the Brooks Range and the Beaufort Sea. There are also some small-scale hard rock gold-mining operations. The Arctic has been little exploited for economic purposes, but, because it contains 8% of the surface of the planet and 15% of the land area, significant resources (both renewable and nonrenewable) may be reasonably assumed to be present.

When the glaciers finally melted they left behind huge piles of boulders, rocks, and debris that contained gold and other important minerals. This gold is what is referred to use glacial gold. Glacial gold is still subject to the same natural erosion that alluvial gold is once it has been deposited. The Arctic region contains several important coal basins, and coal is present in all parts of the Arctic. They are inferred to represent more than 50% of all coal resources of North America. The most popular minerals of Arctic regions are coal, iron ore, zinc, lead, nickel, precious metals, diamonds, and gemstones.

Although it is surmised that the Arctic Ocean area has a certain potential for undiscovered resources, development in the area of extreme cold and sea ice is with difficulties, requiring advanced development technology. Under these circumstances, resources ought to be self-addressed steady over the middle and future from the perspective of continued diversification of resources provides, considering progress in resources development technology in ocean ice regions, cooperative relationships with coastal states, and factors like desires of the personal sector [6].

The Arctic already produces 40% of the world's Palladium (mostly in Russia), 20% of its Diamonds, 15% of Platinum, 11% of Cobalt, 10% of Nickel, 9% of Tungsten, and 8% of Zinc. The "Red Dog mine" lies north of the Arctic Circle in Alaska and is the world's largest Zinc producer, accounting for 5% of global mine production, and the fourth largest Lead producer, responsible for 3% of the world's Lead. There is reason to believe that production could increase further once land becomes more accessible.

The Arctic sits on large deposits of rare earth elements. According to an assessment by the Alaska Division of Geological and Geophysical Surveys, there are more than 150 deposits of rare earth minerals in Alaska alone. As its unmined reserves have become more accessible due to melting ice, the region has become an attractive mining destination.

China has recently shown interest in expanding its presence in Greenland, which has the world's second-largest deposits of rare earth minerals. China already produces 90% of the world's rare earth minerals. Furthermore, a Chinese company is currently in negotiations with London Mining to build an iron-ore mine in Greenland. The strategy seems to be paying off Greenland recently passed the Large-Scale Projects Act, allowing 3000 Chinese workers into the country to build the infrastructure necessary for the mine's operation.

During the investigation of the Arctic region and exploring the North–South route for the movement from Europe to North America, Frobisher got confused that there is a gold-mine present near the Baffin Iceland and after many investigations that mine was proved as a Mirage. But it does not mean that the arctic land and ocean are deficient in minerals. It is the opinion of scientists that beneath the permafrost Arctic, there is a useful and precious reservoir of minerals. Vast deposits of economically valuable mineral resources are available in the Arctic. Significant deposits of phosphate, bauxite, diamonds, iron ore, and gold are in the Arctic region. Deposits of silver, copper, and metallic element conjointly exist within the Arctic; however their extraction is extremely difficult. Deposits of rare-earth metals and other minerals are revealed by the retreating of ice caps and glaciers in Greenland initiating a competition between Europe and China to access this resource [7].

The presence of numerous metallic and nonmetallic minerals, hydrocarbons, gas hydrates, and coal in the Arctic is known for long period. However, due to technical development and decreasing permafrost and sea ice extends these resources are becoming more accessible and getting attention for their exploitation.

5. The stake holders

5.1 Russia

In 1915, Russia became the first nation to drill in the Arctic and has continued to drill in the region since then. As oil and natural gas account for a large portion of Russia's federal budget revenue and exports, Russia has been very much interested in extracting these resources from the region. Russia's share of the oil reserves within the ocean has been calculable to account for half the undiscovered oil within the region. Moreover, 20% of Russia's Gross Domestic Product is generated within the Arctic [8]. The Extreme North or Far North (Russian: Крайний Север, Дальний Север) is a large part of Russia located mainly north of the Arctic Circle and boasting enormous mineral and natural resources. Its total area is about 5,500,000 square kilometers, comprising about one-third of Russia's total area.

5.2 Canada

The United States and Canada jointly banned drilling activity indefinitely in the Arctic, which will be reviewed every 5 years based on a Climate and Marine Science Life Cycle Assessment.

5.3 Denmark

Denmark has expressed interest in Arctic resource exploitation but has stressed the need to do so in a manner that respects the Arctic's nature and environment.

5.4 Norway

Norway encompasses a history of Arctic drillings and continues to specific interest in it. Drilling within the Norwegian ocean bottom began in 1966 and is continuing to be an enormous part of its economic process.

5.5 The United States

In November 2016, citing the need for environmental protection, the US Department of the Interior instituted a ban on drilling in the Beaufort Sea and Chukchi Sea of the Arctic region for the time period between 2017 and 2022 [9].

In view of the foregoing, it may be summarized that Arctic oil and natural gas potential with declining prices, operational risk involved, and disappointing exploratory outcome may receive less attention from petroleum companies. It may result in either a steady production or even a declining trend excepting Russia. Despite the huge quantity of mineral resources in the Arctic region, the required infrastructural support may be inadequate to increase sites and production. Moreover, conserving the Arctic fragile ecosystem and the environment is posing another challenge to all mining activities. Arctic Aggregates: Rock, stone, sand, and gravels are mined throughout the Arctic region for various developmental projects. It is expected that an increased demand of civil construction over the Arctic region will yield more production of such nonliving resources.

It is important to understand that the global supply and demand, global market prices, and political agreements will steer the exploitation and development of the Arctic resources it a large extent. However, the future of Arctic resource extraction is largely technology-driven and to help meet the environmental challenges. Therefore, high economic costs of exploitation and transportation, and high demand for environmental protection are apparently the decisive factors for the exploitation of Arctic resources.

Current climate change and the melting ice in Polar Regions is opening new opportunities to exploit mineral and oil resources, particularly in the Arctic. This emerging situation generates enhanced interest and territorial claims by the Arctic States mainly because of dwindling fossil fuel resources, which is a key to the future of our world economy. Thus, the exploration, development of production, and transport facilities for oil, gas, and mineral resources are increasing throughout the circumpolar region. Receding sea ice cover and permafrost thaw will influence accessibility to mineral and energy resources both on land and in the Continental Shelf in the future.

6. Regional geographic context

The five countries that border the Arctic represent the most important geopolitical players in the region. Each of these nations—Canada, the United States, Russia, Norway, and Denmark (via Greenland)—is entitled to control its own EEZ, which extends 200 nautical miles from its continental shelf, as dictated by the United Nations Convention on the Law of the Sea (UNCLOS). With 157 signatories, UNCLOS, which defines the rights and responsibilities of nations with respect to their use of the world's oceans, is perhaps the most significant international treaty relating to the Arctic. The treaty also establishes guidelines for businesses, the environment, and the management of marine natural resources.

The United States remains one of just a handful of nations that have not yet ratified UNCLOS. As a result, although it abides by the rules of the Convention, it is unable to submit territorial claims to formally define its EEZ in the Arctic. This may place it at a disadvantage, since all the other Arctic nations have moved aggressively to define their EEZs. Because the Arctic is home to what may be the world's largest remaining untapped gas reserves and some of its greatest undeveloped oil reserves, as well as zinc, iron, nickel, and other rare earth metals, those countries that control portions of the Arctic shelf are able to reap significant economic rewards. In December 2014, Denmark became the first country to lay a formal claim to part of the North Pole when it registered a request for about 900,000 square kilometers (350,000 square miles) beyond the coast of Greenland. However, validating and approving claims remains a time-consuming process. With Canada and Russia also likely to submit claims soon, some experts predict that it could take until 2030 or longer before these territorial issues are resolved.

References

[1]   D. Nanda, "India's Arctic Potential", ORF Occasional Paper No. 186, 2019. February 2019.
[2]   S. Chaturvedi, China and India in the 'receding' arctic: rhetoric, routes and resources, Jadavpur Journal of Int. Rel. 17 (2013) 1—46.
[3]   https://geohistory.today/russia-arctic-development-power.
[4]   https://www.asdnyi.com/russia-natural-resources-map.
[5]   https://infogalactic.com/info/Natural_resources_of_the_Arctic.
[6]   http://arctic.or.kr/files/pdf/m4/japan_arctic.pdf.
[7]   https://en.m.wikipedia.org/wiki/Arctic_resources_race.
[8]   https://howlingpixel.com/i-en/Arctic_resources_race) and https://everything.explained.today/Arctic_resources_race/).
[9]   https://everything.explained.today/Arctic_resources_race .

Chapter 9

Arctic Ocean circulation

Among the world's five major oceans, the Arctic Ocean is the smallest, shallowest, and coldest [1]. Although the International Hydrographic Organization (IHO) recognizes it as an ocean (Fig. 9.1A), yet according to some oceanographers the Arctic is considered as the Sea. It is also seen as the northernmost part of the World Ocean. Surprisingly, sometimes the Arctic is also considered as an estuary of the Atlantic Ocean [2,3]. Arctic Ocean encompasses the areas even up to 57 degrees N as lineated by the IHO. It includes Baffin Bay, Barents Sea, Beaufort Sea, Chukchi Sea, East Siberian Sea, Greenland Sea, Hudson Bay, Hudson Strait, Kara Sea, Laptev Sea, White Sea, and other tributary bodies of water. It is connected to the Pacific Ocean by the Bering Strait and to the Atlantic Ocean through the Greenland Sea and Labrador Sea [1]. Fig. 9.1B and C shows Arctic Ocean bottom features depicting various oceanic ridges and basins.

The Arctic (north pole) region lies in the middle of the Northern Hemisphere, where the Arctic Ocean occupies a roughly circular basin and covers an area of about 14,056,000 km^2 (equivalent to the size of Antarctica) [4,5] with about 45,390 km long coastline [4,6]. The Arctic Ocean has a total volume of 18.07×10^6 km^3, equal to about 1.3% of the World Ocean. It is surrounded by Eurasia and North America Greenland, and several islands. It is partly covered by sea ice throughout the year and almost completely in winter. Depending upon the seasonal changes (melting/freezing) of the ice cover over the Arctic Ocean, its surface temperature and salinity vary [7]. Primarily due to low evaporation, heavy freshwater inflow from rivers and streams, and limited connection and outflow to surrounding oceanic waters with higher salinities, the Arctic Ocean records the lowest salinity among the five major oceans [7]. Countries bordering the Arctic Ocean are Russia, Norway, Iceland, Greenland (territory of the Kingdom of Denmark), Canada, and the United States.

1. Currents and circulation

The Arctic Ocean exhibits a uniquely complex system of water flow primarily due to its relative isolation from other oceans. Ocean currents are responsible for the movement of water from one ocean region to another, due to gravity, wind, and the rising and sinking of water in different parts of the world, owing

The Arctic. https://doi.org/10.1016/B978-0-12-823735-9.00006-0

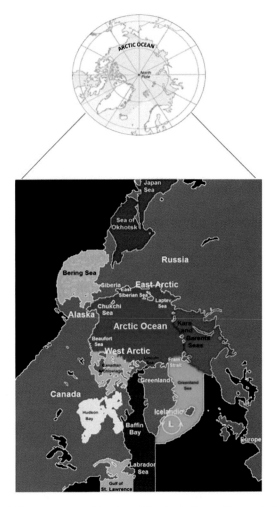

FIGURE 9.1A Different sectors and borders of the Arctic Ocean. *Source: A. J. Luis, personal communication.*

to varying temperatures [8]. It is a known fact that the temperature affects the density of water. Accordingly, due to warming, it expands and becomes less dense, causing it to rise to the surface. Such warm water evaporates quickly, which tends to increase its salt concentration. Once the salt concentration increases, it leads to the surface water's density, and accordingly dense water sinks, causing the upwelling of cooler waters, which is again exposed to the sun heat (warmed up) to continue the cycle (Fig. 9.2A and B).

In large parts of the Arctic Ocean, the top layer (about 50 m) is of lower salinity and lower temperature than the rest. Between this lower salinity layer

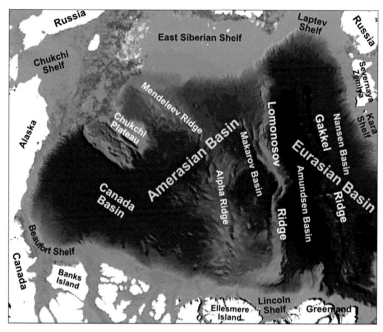

FIGURE 9.1B Arctic Ocean bottom features. *Source: https://upload.wikimedia.org/wikipedia/commons/d/d4/Arctic_Ocean_bathymetric_features.png.*

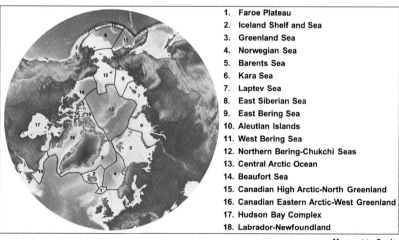

1. Faroe Plateau
2. Iceland Shelf and Sea
3. Greenland Sea
4. Norwegian Sea
5. Barents Sea
6. Kara Sea
7. Laptev Sea
8. East Siberian Sea
9. East Bering Sea
10. Aleutian Islands
11. West Bering Sea
12. Northern Bering-Chukchi Seas
13. Central Arctic Ocean
14. Beaufort Sea
15. Canadian High Arctic-North Greenland
16. Canadian Eastern Arctic-West Greenland
17. Hudson Bay Complex
18. Labrador-Newfoundland

Map not to Scale

FIGURE 9.1C The 18 major Marine Ecosystems of the Arctic Ocean. *Source: https://www.pame.is/projects/ecosystem-approach/arctic-large-marine-ecosystems-lme-s.*

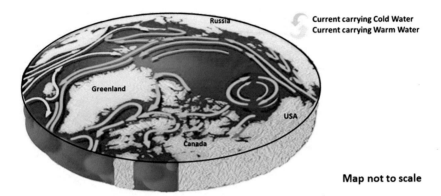

FIGURE 9.2A Warm and cool currents in the Arctic Ocean (modified). *Source: http://www. aquatic.uoguelph.ca/oceans/ArticOceanWeb/Currents/frontpagecur.htm.*

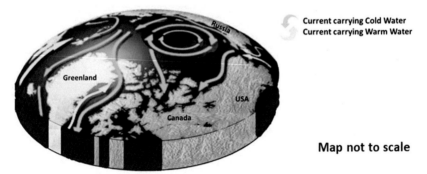

FIGURE 9.2B Warm and cool currents in the Arctic Ocean (modified). *Source: http://www. aquatic.uoguelph.ca/oceans/ArticOceanWeb/Currents/frontpagecur.htm.*

and the bulk of the ocean lies the so-called halocline, in which both salinity and temperature rise with increasing depth. Fig. 9.3 shows various basin-wise density structure of the upper 1200 m in the Arctic Ocean.

A gyre is, generally, the circulation of the surface waters of the Arctic Ocean in a large clockwise rotational pattern moving from east to west around the polar ice cap. The movement of ocean currents in a clockwise direction in the Northern Hemisphere, and in an anticlockwise direction in the Southern Hemisphere is governed and controlled by the Coriolis forces associated with the Earth's rotation. The lack of movement in the deep waters has caused a stagnant pool of very cold water to accumulate at the bottom of the Arctic Basin because only surface waters are exchanged via ocean currents as the exchange of very deep waters is prevented due to high submarine ridges.

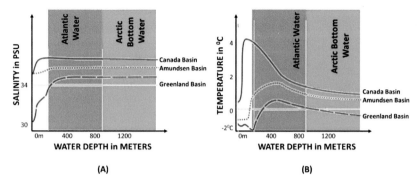

FIGURE 9.3 Density structure of the upper 1200 m in the Arctic Ocean for different basins (modified). *Source: https://en.wikipedia.org/wiki/Arctic_Ocean#/media/File:Temperature_and_salinity_profiles_in_the_Arctic_Ocean.svg.*

The Arctic surface waters extend to a depth of about 46 m and are far less salty than the waters below. The heating of the Arctic Ocean is not only because of solar incoming radiations but also due to the heat that comes from the south with ocean currents and airstreams.

Although the currents change their name as they move north, nevertheless they are all part of an extension of the Gulf Stream. The North Atlantic Current, a branch of the Gulf Stream flows along the coast of Norway and continues all the way to the Arctic Ocean and known as the West Spitsbergen Current carrying almost 60% of the water entering the Arctic Ocean (Fig. 9.4).

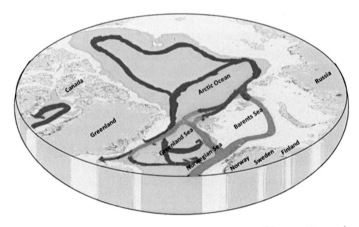

Map not to scale

FIGURE 9.4 Warm and cool currents in the Arctic Ocean (modified). *Source: http://www.arcticsystem.no/en/outsideworld/oceancurrents/.*

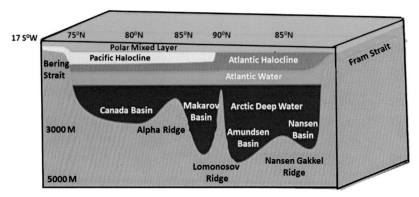

FIGURE 9.5 Vertical section of major water mass in the Arctic Ocean (modified). *Source: http:// www.homepages.ed.ac.uk/shs/Climatechange/Geo-politics/Ice/Arctic_Ocean.htm.*

Interestingly, the Bering Strait and the big Russian and Canadian rivers, carry a good amount of freshwater leading to less salinity of the topmost 45 m of the Arctic Ocean. Between Greenland and Svalbard, the Fram Strait is the major arterial route for seawater to and from the polar area. Although warm Atlantic Ocean water flows north along the west coast of Svalbard, and cold water flows south along the coast of East Greenland, and out into the Barents Sea along the east side of Svalbard. Fig. 9.5 demonstrates the major water mass in the Arctic Ocean along a vertical section of different water masses from Bering Strait over the geographic North Pole to Fram Strait. Deeper water masses are denser than the layers above.

The cold, relatively fresh, Arctic water meets the warm, saline Atlantic water in the Barents Sea, and is called the polar front with varying geographical position.

The predominant wind direction over the Arctic Ocean rotates the surface water in a large circle, called The Beaufort Gyre, caused due to strong winds over the Arctic Ocean (Fig. 9.6) that rotates the surface water in a large circle (Gyre). Leading to slow polar ice movement in the same direction, too. It takes about 2 years for the Sea ice that lies close to the center of the gyre but takes a longer period of 7−8 years for the ice of the furthest from the center requires to complete a full 360 degrees circle. The Transpolar Drift consists of wind and ocean currents that cross the entire ocean from Siberia to exit through the Fram Strait is another dominating surface circulation [8].

Within the Barents Sea, a smaller gyre flows in a counter-clockwise direction, with warmer Atlantic waters from the south mixing with colder Arctic waters in the north. This counter-clockwise movement occurs in response to a counter-clockwise wind system that moves across the Barents Sea. The Barents Sea current is generally weak and variable, as are the equivalent gyres that occur in the Norwegian and Greenland Seas. An

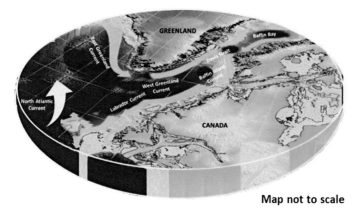

Map not to scale

FIGURE 9.6 Formation of Beaufort Gyre, Barents Gyre, and Labrador Current (modified). *Source: https://en.wikipedia.org/wiki/Labrador_Current#/media/File:LabradorCurrentus-coast guard.jpg.*

additional current, the West Greenland Current, dominates the waters off the coast of Greenland, south of Davis Strait. As the West Greenland current approaches the Davis Strait, it joins the Labrador Current, and then continues northward into the Baffin Bay where it cools down dramatically.

About 90% of the sun's energy is reflected due to the vast snow and ice that cover the Arctic. And thus, constant heat is lost from the Arctic. Arctic Ocean currents are not only influenced by wind direction, but also by the inflow from the rivers such as Mackenzie River in Canada and the Ob, Yenisey, and Lena Rivers in Siberia, that discharge into it (Fig. 9.7A). The North Atlantic Current

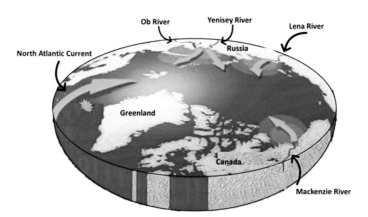

Map not to scale

FIGURE 9.7A Areas of inflow for the Arctic Ocean (modified). *Source: http://www.aquatic. uoguelph.ca/oceans/ArticOceanWeb/Currents/inflow.htm.*

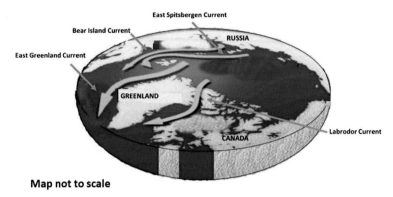

Map not to scale

FIGURE 9.7B Outflow from the Arctic Ocean (modified). *Source: http://www.aquatic.uoguelph. ca/oceans/ArticOceanWeb/Currents/outflow.htm.*

brings warmer water from the Atlantic Ocean, while the Bering Strait Outflow from the Arctic Ocean also moves some water into the Arctic Ocean from the Bering Sea and the Pacific Ocean [8].

To maintain a constant sea level, the influx of water from various waterways must be matched by an outflow. Most of the outflow is via currents (Fig. 9.7B). Water flows from the Arctic Ocean into the Pacific and Atlantic Oceans, as well as into several surrounding seas. By far, the greatest volume of water leaves the Arctic Ocean through the passage between Greenland and Spitsbergen. This enormous outflow creates the cold East Greenland Current in the Greenland Sea. A weaker cold current, the Labrador Current, flows south through Smith Sound and Baffin Bay. A considerable amount of water also flows from the Arctic Ocean into the northern Barents Sea forming the East Spitsbergen and Bear Island Currents [8−10].

This complex circulation system in the Arctic impacts the entire food web. A dramatic reduction in sea ice cover and a weakening of the Beaufort Gyre circulation system owing to ongoing climate change. Such unprecedented changes in the Arctic Ocean circulations not only affect locals but also people living far away.

2. Global conveyor belt

The global conveyor belt (Fig. 9.8A and B) is a system of ocean currents that transport water around the world. While wind primarily propels surface currents, deep currents are driven by differences in water densities in a process called thermohaline circulation. The ocean circulation conveyor belt helps balance climate. The global ocean conveyor belt is a constantly moving system of deep-ocean circulation driven by temperature and salinity. This motion is caused by a combination of thermohaline currents (thermo = temperature; haline = salinity) in the deep ocean and wind-driven currents on the

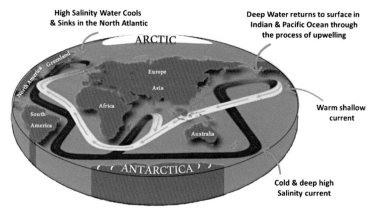

Map not to scale

FIGURE 9.8A Global conveyor belt (thermohaline circulation). *Source: https://nsidc.org/about/overview.*

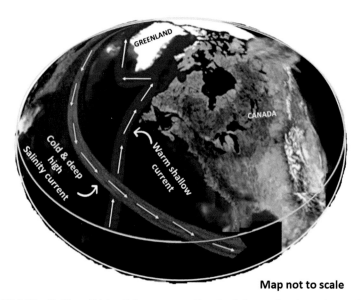

Map not to scale

FIGURE 9.8B Sinking of high salinity water near Greenland after cooling down. *Source: https://oceanservice.noaa.gov/facts/conveyor.html.*

surface [11]. Dependent upon the conjoint effect of temperature and salinity thermohaline circulation in the Pacific. Thermohaline circulation is a part of the large-scale ocean circulation that is driven by global density gradients created by surface heat and freshwater fluxes. The thermohaline circulation is sometimes called the ocean conveyor belt, the great ocean conveyor, or the global conveyor belt.

There are five gyres, the North Atlantic Gyre, the South Atlantic Gyre, the North Pacific Gyre, the South Pacific Gyre, and the Indian Ocean Gyre, that have a significant impact on the ocean. The big five help drive the so-called oceanic conveyor belt that helps circulate ocean waters around the globe.

As a part of the ocean conveyor belt, warm water from the tropical Atlantic moves poleward near the surface where it gives up some of its heat to the atmosphere. This process partially moderates the cold temperatures at higher latitudes. Thermohaline circulation drives a global-scale system of currents called the "global conveyor belt." The conveyor belt begins on the surface of the ocean near the pole in the North Atlantic where the water is chilled by arctic temperatures.

The global conveyor belt moves much more slowly than surface currents a few centimeters per second, compared to tens or hundreds of centimeters per second. Scientists estimate that it takes one section of the belt 1000 years to complete one full circuit of the globe. If ocean currents were to stop, the climate could change quite significantly, particularly in Europe and countries in the North Atlantic. In these countries, temperatures would drop, affecting humans as well as plants and animals. Shutdown of circulation pattern could be disastrous, researchers say. If global warming shuts down the thermohaline circulation in the North Atlantic Ocean, the result could be catastrophic climate change. Between Greenland and Norway, the water cools, sinks into the deep ocean, and begins flowing back to the south.

References

[1] M. Pidwirny, Introduction to the Oceans, 2006. Archived from the original on 9 December 2006. Retrieved 7 December 2006, www.physicalgeography.net.

[2] M. Tomczak, J.S. Godfrey, Regional Oceanography: An Introduction, second ed., Daya Publishing House, Delhi, 2003, ISBN 978-81-7035-306-5. Archived from the original on 30 June 2007. Retrieved 22 April 2006.

[3] Arctic Ocean — Encyclopedia Britannica. Retrieved 2 July 2012. As an Approximation, the Arctic Ocean May Be Regarded as an Estuary of the Atlantic Ocean.

[4] J.W. Wright (Ed.), The New York Times Almanac, 2007 ed., Penguin Books, New York, 2006, ISBN 978-0-14-303820-7, p. 455.

[5] "Oceans of the World" (PDF). rst2.Edu. Archived from the original(PDF) on 19 July 2011. Retrieved 28 October 2010.

[6] Arctic Ocean Fast Facts. wwf.pandora.org (World Wildlife Foundation). Archived from the original on 29 October 2010. Retrieved 28 October 2010.

[7] K. Aagaard, R.A. Woodgate, Some thoughts on the freezing and melting of sea ice and their effects on the ocean, in: Polar Science Center, Applied Physics Laboratory University of Washington, January 2001. (Accessed 7 December 2006).

[8] http://www.aquatic.uoguelph.ca/oceans/ArticOceanWeb/Currents/frontpagecur.htm.

[9] http://www.arcticsystem.no/en/outsideworld/oceancurrents/#:~:text=One%20branch%20of%20the%20Gulf,Arctic%20Ocean%20comes%20with%20this.

[10] https://www.whoi.edu/know-your-ocean/ocean-topics/polar-research/arctic-ocean-circulation/.

[11] https://oceanservice.noaa.gov/facts/conveyor.html.

Chapter 10

The Arctic: Ocean of the future

The human population on the earth is not equal in both hemispheres. The 9/10 part of the total human population resides in the Northern hemisphere and 1/10 in the Southern hemisphere. Apart from China and India, the capitals of the countries of the Northern Hemisphere are closer to Arctic Circle in comparison to the Equator. The revolutionary development in transport, especially marine and aviation, has confirmed the forecast made by Vilhjalmur Stefansson that "in future the Arctic will become so important as is Mediterranean Sea in present" to some extent. Stefansson (1879–1962) was an interesting Arctic explorer. He traveled the Arctic region many times and had suggested that without any external help for food and fuel, no man can survive in the extreme climate conditions of the Arctic. He had set up a Scientific Center in the ice land in 1918.

In his travel details, Pythus had mentioned that "in the North of British Islands on a distance of six days sea travel, there is a sea which is always frozen." His statement was true. He was talking about Artic. At the distant North of the earth, there is a sea which is the Arctic which we all today recognize. Due to various reasons, geographers do not consider it an ocean. They consider it as the expansion of the Pacific and Atlantic sea. It relates to the Pacific Ocean through Bering Strait in the East. In the West, it relates to the Atlantic sea through Barents, Greenland (which are the part of Arctic Ocean) and Norwegian seas by Fram Strait. The geographic North Pole is situated in the middle of the Arctic Ocean. Though, approximately 70% of the ocean remains covered by ice for the whole year, despite the Arctic Ocean is strategically an important ocean. It provides the shortest route between Russia and the United States of America. Therefore, during the Second World War, both countries had set up various Observation and Scientific Centers at the Arctic. These stations not only kept working after the end of the war, but they increased in number and were equipped with the latest technology and facilities as time passed by.

In 1960, Russia conducted atomic experiments in Novaya Jambalaya Islands located in the Arctic Ocean. Arctic was recognized as an oil passage by the oil tanker S.S. Manhattan by sailing through it in 1969.

The Arctic. https://doi.org/10.1016/B978-0-12-823735-9.00013-8

1. Ecotourism

Ecotourism is also becoming an important activity in the Arctic region, with 250 cruise ships venturing across the Arctic in 2008, compared with just 50 in 2004, according to statistics from the State University of New York's Maritime College. Japan's Crystal Cruises planned to take 900 passengers on a 32-day cruise through the Northwest Passage in 2016, making it the largest and most luxurious ship to ever make the trip. However, the prospect of more Arctic ecotourism is raising growing safety and environmental concerns. In 2010, a cruise ship ran aground in the Canadian Arctic, even though traveling conditions were perfect. A coast guard icebreaker was able to rescue the passengers and crew in enough time, but the result could have been different had the weather conditions been dangerous. Canada is preparing for the expected uptick in Arctic-bound passenger ships using tactics such as Operation Nanook, a joint training exercise run by the Canadian Armed Forces and Coast Guard, which in 2014 simulated the rescue of a 50-passenger cruise ship. The Polar Code now sets international safety and environmental specifications for cruise ships in the Arctic but does not include construction requirements for passenger ships operating in polar waters.

2. Expansion

Arctic Ocean is located at 70 degrees North of North Latitude and it includes the Polar Sea (popularly known as Arctic Sea) and connected Greenland, Norwegian, Barents, Kara and Laptev Sea, Baffin Bay, and water bodies of Canadian Islands also. Its total area is 1,39,86,000 sq. km. If Hudson Bay and Hudson Strait is also included, its area increases by 4,80,000 sq. km. Despite this, it is the shortest Ocean in the world. Geographically, the Arctic spans the Arctic Ocean and covers land areas in parts of Canada, Finland, Greenland, Iceland, Norway, Russia, Sweden, and the United States (Alaska). Eurasia, North America, Canada, and Alaska are situated around the Arctic Ocean. The quality of its water is different from the water of the Pacific and the Atlantic Ocean but like the small seas included in it.

The maximum width of the ocean is 4233 km between Alaska and Norway and minimum 1900 km between Greenland and Taimyr Island (Russia). Its average depth is 1330 m. Like other oceans, it also has straits, ditches, mountains, and continental shelves in it. The width of its continental shelves is not equal on all coasts and it varies. The width of the shelves is 100−200 km between the North coasts of Alaska and Greenland, but the Siberian and Chukki Shelves are 500−1700 km wide. The depth of continental shelves is 150 km. In all, the area of its continental shelves is half that of its total area. There are many drenches in this continental shelf. The depth of some parts of the Arctic Ocean ranges from 3050 to 3660 m, and its deepest region is 5450 m. This region is near Svalbard Islands located between Norway and the North Pole. The depth of the Arctic Sea near the North Pole is 4270 m.

In the middle of the Arctic Ocean, there is a circular basin that is 1100 km wide and 2250 km long. It divides the Arctic into two equal parts. The ridge Lomonosov is submerged in it. The Lomonosov Ridge is a 1800 km long ridge, which divides the Arctic Ocean into two major basins: the Eurasia Basin and the Amerasia Basin. The Amerasia Basin is more than 4000 m deep, whereas its Eurasian counterpart measures 3400 m. The ridge itself rises 1800−3400 m from the basin floor (Fig. 9.1B).

This ridge was explored by Russian scientists in 1948−49. The ridge is stretched from Elsmere Island to New Siberian Island. It is 1800 km long, and it passes near the North Pole. It is 3000-m-high from the sea level and is like the structure of the mountains of Siberia. As it was explored by Russian scientists, it has claimed its ownership on the area near the ridge, but other countries did not agree to it.

In addition to the Lomonosov Ridge, two other ridges—Alpha and Mid Atlantic Oceanic ridge—are submerged in the Arctic Ocean. Svalbard, Franz Joseph Island, and Severna Jemalya islands are the tops of these ridges that emerged out of the sea.

3. Water

The temperature and salinity of the Arctic Ocean are not similar in all regions. Therefore, the Oceanographers divide the water bodies into four parts, that is, Arctic upper layer water, water of the Pacific Ocean, Water of Atlantic Ocean, and Arctic sea level water. The salinity of the upper layer water is the lowest (30.33%), and the salinity is the most in bottom level water (34.9%). The saline water is up to 50-m depth. During summers, its temperature remains $-1°C$ and during summer approx. $-2°C$. The water of the Pacific Ocean remains between the Lomonosov Ridge and the Pacific Ocean only. It is available up to 125-m depth. Its temperature remains $-1°C$. On the contrary, the water of the Atlantic Ocean is also found under the layer of the Pacific Ocean and its salinity is comparatively less. It is found till 850-m depth approximately and its temperature is $2°C$. The salinity of the lower-level water is the most. Its temperature remains $-1°C$.

There is a narrow strait between the Arctic Ocean and the Pacific Ocean. The exchange of water between these two oceans does occur but in a very less quantity. The exchange of water occurs with the Atlantic Ocean through the Greenland Sea. Small ridges create hurdles in the exchange of water. Therefore, the temperate of upper-level water of these Oceans mix with each other but the lower-level water of deep regions cannot mix with each other. Therefore, a vast reservoir of cold water gets created in the lower level of the Arctic Ocean. Three big rivers of Siberia—Lena, Yenisei, and Ob—and Canada's Mackenzie and other small rivers fall in the Arctic Ocean. These are the sources of freshwater for the Arctic Ocean. Moreover, 90% of freshwater is provided by Siberian and 10% from Canadian Rivers. Though the Arctic gets very less rainfall but due to its cold climate, the rate of evaporation is very less. With this background, drifting of its water to other parts through the Greenland Sea is a significant event. This water forms the East Greenland stream. It is an extremely cold-water stream. In addition to this, another stream also flows from the Baffin Bay. Popular as Labrador Stream, it is also a narrow water stream. Another stream flows from the Baring strait toward the Pacific Ocean.

4. The ice

More than 70% of the Arctic Ocean is composed of a drift ice pack that can be up to nine feet thick during winters floating above the water. The concentration of the ice is maximum in the Siberian and North American regions. The layer of the ice is called Polar Ice Cap. The ice does not freeze on the West Coast of Norway due to temperate stream—North Atlantic Stream and Norwegian Stream flow through it. Much of the Arctic Ocean is composed of drifting icebergs, ice islands, and large ice packs. The layer of the ice on the ocean can be up to 3 m thick in the ocean and 2 m in coastal areas. The region near the North Pole is covered by ice for the whole year. It melts near the coasts during summers. In the summers, the ice pack (Fig. 10.1) is replaced mainly by open water that is often dotted with icebergs that are formed when ice is broken by wind, sea, swells, currents, and tides. Floating large ice packs keep striking with each other due to the streams of water and because of flowing winds it makes the sound like the thundering of clouds.

During summers when most of the coastal areas become ice free the movement of ships start, and they keep sailing regularly on Northeast and Northwest Passages. The waterways are located near the coasts. Three types of ice are found in seas of the Arctic region: Sea Ice, Ice Bergs (Fig. 10.2), and Ice Islands.

FIGURE 10.1 Floating ice pack in the Arctic Ocean. *Photo Courtesy Shabnam Choudhary.*

FIGURE 10.2 Iceberg in Arctic. *Source: https://upload.wikimedia.org/wikipedia/commons/3/3d/ Iceberg_in_the_Arctic_with_its_underside_exposed.jpg.*

The Sea Ice is formed in the sea by itself due to the lowering of temperature. The ice packs, icebergs, and Ice Islands are formed due to the falling of glaciers into the Arctic Sea. Generally, the Sea Ice and Ice Islands are formed at high altitudes where the temperature of the water is quite low and therefore they do not melt easily. The formation of Sea Ice depends on many factors viz. temperature of the environment, salinity of the seawater, freezing period, and depth of the sea. The salinity is an important factor because the freezing point is inversely proportional to salinity. In other words, the more is the salinity; the ice will melt at that minimum temperature.

The depth of the ocean is another factor for the formation of ice because the depth controls the movement of water in the sea. Sometimes the upwelling activity lowers the ice formation activity of sea ice as due to upwelling the comparatively temperate water of the sea level keeps moving toward the upper layer of sea and the upper layer water moves to the sea level. The water needs to be stable and calm for the formation of ice. The ice formation either lowers or stops where the water movement is high.

While converting from water to ice, the heat consumed while conversion of one cubic meter water is enough to increase the temperature of 3118 cubic meter air up to 10°C. From this, we get the idea about the vast hidden heat available in seawater. The freshwater freezes at a higher temperature than the saline water. That is why when the seawater converts into ice all the salts get separated from it. In this way, the sea ice should not contain salt, but it is not so. While converting into ice, some quantity of salt gets stick to it. That is why the ice becomes partially saline up to 20%. The more are the pores on the ice, the more light is the ice and its density is less. The seas with more salinity have lower hidden heat. The characteristics of the ice depend on the method of its formation. If the environment and sea are calm the ice is crystal, clear. If the layer is disturbed the ice is smooth with air particles inside it. If there is little movement in the sea during ice formation, the circular pieces are formed which keep floating on the surface of the water. When the water freezes inside the coastal areas of the bay or the seawater inside the land then the ice remains stable for long on its place for many years but, generally, it breaks and gets separated from the coastal area. Such ice is called Fast Ice. It keeps sticking with the coasts.

5. Icepacks

When the glaciers come drifting toward the coasts and their pieces fall into the sea, these pieces form Icepacks (Fig. 10.3). They can be very big. They are irregular in size. Many such icepacks can be seen floating on the Arctic Ocean which is hundreds of meters wide and tall. It is believed that most of the Icepacks of the Arctic Ocean are formed by the breaking of large continental Glaciers of Greenland. Approximately 10,000 icepacks are formed during a year. Out of these ice packs, three-fourth icepacks flow toward the south by Labrador Stream.

6. Ice Islands

Ice Islands are unique and typical structures that have perplexed many initial explorers of the Arctic. They are those pieces of Glaciers that are formed by the falling of Ellesmere continental glaciers into the ocean. They can be 60-m-thick and 530-m-wide in size. They are formed on the land surface; therefore, pieces of rocks, sand, and mud are merged with them. These islands had created doubt in the mind of initial explorers that land formations like "Cracker land" are available in the distant north. Now, it is clear that no such landforms exist there.

FIGURE 10.3 Arctic ice pack. *Source: https://upload.wikimedia.org/wikipedia/commons/6/68/ Arctic_ice.jpg.*

The significance of Ice Island was recognized in the last part of the decade during 1940 and during that period many scientific experimentation centers were being established there. For the first time, a famous scientific center T-3 was established on "Flatchers" Ice Island. The navy of the United States of America and 24 staff members of the Air Force worked there. It is said that such centers were established in other Ice Islands and many people are working with them.

Most of the ice packs are formed during the spring and summer seasons and especially during May when the rate of melting of glaciers increases due to high temperatures. These ice packs and Ice Islands sometimes reach up to Gulf Stream and most of them melt before reaching Grand Banks near New Found land. Sometimes they come on the sea route between North America and Europe and create hazards for the ships. On April 15, 1912, the Titanic sank after hitting a large Ice Island. In this mishap, 1490 lives were lost.

7. The biology

The Arctic is a very cold ocean with minimum or no sunlight sun during the winters. The days are comparatively longer and bright during the summer season. Earlier people were of the view that no life is possible in this cold ocean region but now we have come to know that the Arctic is a habitat for various species of animals ranging from single-cell Plankton to large animals like Krill, Copepod, Seal, Walrus, Sea Lion, and Whale. According to the environment, all these animals have adapted themselves to the Arctic. Some of the animals have certain elements in their bodies that help avoid freezing of their blood during the extreme cold of the Arctic and some animals are saved by the thick layer of fat and fur on their bodies.

8. Single-cell animals

Planktons of the plant as well as animal origin are found in the Arctic Ocean. Planktons are a diverse group of organisms that live in the water column and cannot swim against a current. The planktons that get energy from photosynthesis are called phytoplankton. They provide a crucial source of food to many species of aquatic animals such as fish and whales. Like Phytoplankton, Zooplankton cannot get food themselves. They have weak swimming capability. The zooplanktons are also food for many aquatic animals like phytoplankton. There are many species of zooplankton.

One of the most valuable resources in the Arctic is debatably the trade routes. Currently, there are three shipping passages in operation viz: the Northeast Passage (NEP), the Northwest Passage (NWP), and the Trans-Polar Passage (TPP) [1]. While these passages are currently in use, other passages that previously were inaccessible are now becoming accessible due to climate change and sea ice melting. Moreover, 80% of the most industrialized nations are fostered by the shorter trade routes of the Arctic Ocean. The consequences of these shorter trade routes are less fuel consumption, less carbon emissions, and faster transportation of goods.

9. Undiscovered resources

Research on the undiscovered resources was conducted by United States Geological Survey (USGS) at the north of the Arctic Circle and revealed that at least 50 million barrels of natural gas and oil is having accumulated underneath the arctic seabed [1]. The USGS estimates that over 87% of the Arctic's oil and natural gas resource is in seven Arctic basin provinces: Amerasian Basin, Arctic Alaska Basin, East Barents Basin, East Greenland Rift Basin, West Greenland-East Canada Basin, West Siberian Basin, and the Yenisey-Khatanga Basin [2,3].

10. New opportunities

New trade route potential was explored by Shipping companies and countries. Another significant economic driver for the region is the abundant presence of energy both onshore and within the exclusive economic zones of the five Arctic coastal nations. On the limits of the continental shelf, several countries have already submitted their claims to further extend claims under the Commission on the Limits of the Continental Shelf [4]. This includes several overlapping claims, to include the Lomonosov Ridge and the North Pole. The United States remains the only nation that has not ratified the United Nations Convention on the Law of the Sea (UNCLOS); it appears that all Arctic nations will submit claims in accordance with UNCLOS [5]. While oil and gas reserves are still unknown, it is estimated that the Arctic may hold nearly one-third of the world's natural gas and 13% of global oil reserves. Yet, the costs of exploring, developing, and extracting these resources are very high given the harsh environment, limited infrastructure, and difficulties posed. Given the current market prices, there is limited interest in pursuing these reserves in North America, though Norway and Russia are continuing development in the Barents and Kara Seas. Overall production of Arctic energy reserves will likely remain limited soon unless the price of oil climbs significantly. Other sources of oil and gas to include shale and using newer technology in older fields will continue to remain a more economical option.

Mineral resources are also found in vast quantities throughout the Arctic, with all Arctic nations except Iceland possessing significant mineral deposits. While some new deposits are being revealed as the ice cover melts, it is likely that development in the near time will continue to focus on existing mines. It is predicted that infrastructure to these mines and areas will steadily be improved to permit future access. Climate changes are also likely to result in increased fishing in the Arctic. Therefore, Arctic coastal states need to work together in regulating and monitoring fishing activity. Likewise, the regulation on tourism in Arctic water and the establishment of clear safety and emergency response protocols will require cooperation from the Arctic states as the numbers of tourists rise in the future.

The other area of potential disputes relates to the opening of new shipping routes owing to the melting Arctic ice. Canada, for one, holds the Northwest Passage to be falling within its territory of internal waters; the United States, for its part, regards the passage as part of international waters [6]. The stakes are significant as the new shipping routes will likely give economic returns with the shortening of journey time and the reduction of costs. In 2018, cargo shipments through the section of the Arctic sea route near Russia reached 15 million tonnes, over five times higher than the number in 2013 [7]. Estimates suggest that by 2025, over 60 million tonnes of energy resources will be transported via the Northern Sea Route, including coal and Yamal Energy project-produced LNG [8]. The number is likely to increase with the ease of navigability triggered by ice-melt. Fig. 10.4 shows a potential Arctic sea route excepted due to ice-melting.

These new routes may yet prove to be more secure than conventional routes that are affected with problems such as piracy and terrorism [9].

Moreover, these coastal states might act to increase their military presence in the region given their insecurities, resulting in a tenser and less peaceful environment. Indeed, there are still remnants of the militarization of the Arctic that was witnessed in the Cold War [10]. Today, what could compound the matter is that the Arctic does not have any regulations in place against military use by any of the littorals. Also, military and security matters fall outside the ambit of the Arctic Council [11].

11. Future excessive release of methane

Gas hydrates stability zone is especially sensitive to climate change and thus ocean bottom temperatures, geothermal gradient, salinity of the water, and composition of gas. The Arctic Ocean most likely has a larger gas hydrate stability zone compared to other oceans owing to its cold water and low geothermal gradients. It is estimated that a vast quantity of methane is trapped in oceanic hydrate deposits, which are likely to be dissociated due to the expected rise in the ocean temperature which in turn carry the potential release of large amounts of methane into the atmosphere [1,12]. In the future, the Arctic should be considered a critical area in a warming climate because massive amounts of methane are currently locked as gas hydrates in ocean sediments and in permafrost that could be released [1,5]. It hints at the possibility of a previously undiscovered, stable reservoir for methane that is "locked" away from the atmosphere, be released in the future, and thus impact global climate change.

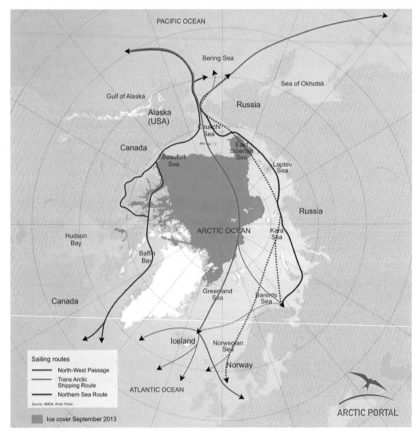

FIGURE 10.4 Potential Arctic sea route. *Source: https://arcticportal.org/images/news/2019/Arctic_Sailing_Routes_hd.jpg.*

References

[1]   https://en.wikipedia.org/wiki/Arctic_resources_race.

[2]   https://www.eia.gov/todayinenergy/detail.php?id=4650.

[3]   http://www.earthbyte.org/Resources/ICONS/index.html.

[4]   https://geology.com/articles/arctic-oil-and-gas/.

[5]   https://www.iucn.org/downloads/arctic_law_policy.

[6]   https://en.wikipedia.org/wiki/Arctic_policy_of_Russia.

[7]   U.K. Sinha, A. Gupta, The arctic and India: strategic awareness and scientific engagement, Strateg. Anal. 38 (6) (2014) 8740.

[8]   RT News, Cargo Shipments along Russia's Arctic Sea Route Reach 15 Million Tons, RT.Com, 2018. November 23, 2018.

[9]   The National, LNG shippers set to gain as Arctic sea routes open up, The National, 2018. September 2, 2018.

[10]   S. Chaturvedi, China and India in the 'Receding' arctic: rhetoric, routes and resources, Jadavpur J. Int. Rel. 17 (2013) 1−51.

[11]   U.K. Sinha, A. Gupta, The arctic and India: strategic awareness and scientific engagement, Strateg. Anal. 38 (6) (2014) 875.

[12]   Arctic Council, The Arctic Council: A Backgrounder, about Us, 2018. Accessed November 8, 2018.

Chapter 11

Arctic governance

The Arctic region comprises the Arctic Ocean and the adjacent seas along with parts of several countries, including the United States (northern parts of Alaska), Russia (northern parts of Siberia's mainland), Canada, Greenland (Denmark), Finland, Norway (Svalbard archipelago), Sweden, and Iceland (Fig. 11.1). The Arctic cryosphere is either receding or crumbling due to ongoing climate change and global warming. The rising temperature is not only affecting the biodiversity and human population but also the geographical features of the Arctic region. It gives rise to new social, economic, security, and political dynamics in the region. Undoubtedly, a new Arctic needs new rules. As climate change causes the Arctic's ice to melt and new areas to open, the region is facing unprecedented changes and serious threats. The ecosystems of the Arctic transcend political boundaries, making collaboration among Arctic states essential. The need to work together is intensified by the sparse population and limited resources of the region. That is clearly visible in the Arctic, a region undergoing profound environmental, economic, and social change, where new actors are coming to the fore and new regulatory frameworks are emerging. No single researcher or a single study can give justice to this richness encompassed within the concept of "Arctic governance."

Arctic governance is a multifaceted, multitier, multilevel, and multidimensional concept that amalgamates both public and private governance. It brings synergy among regulations, good practices, institutions, and set targets such as sustainability. It is a dynamic notion and encompasses an objective approach while keeping its goals and handling any issue or policy related to the Arctic and its stakeholders. Arctic Council being the pivot of the Steering of the entire gamut of activities in the Arctic region, is duty bound to ensure that all Arctic states are taking appropriate actions to fulfill their responsibilities as the primary stewards of the region.

With regards to Arctic-related issues, Arctic and Non-Arctic States have different rights, interests, and specific concerns ensuring peace, stability, and sustainable development in the Arctic region.

In a region of growing global importance, mutually beneficial cooperative partnerships that promote and enhance these interests will surely be the most appropriate way forward. Cooperation between the Arctic and non-Arctic

The Arctic. https://doi.org/10.1016/B978-0-12-823735-9.00015-1

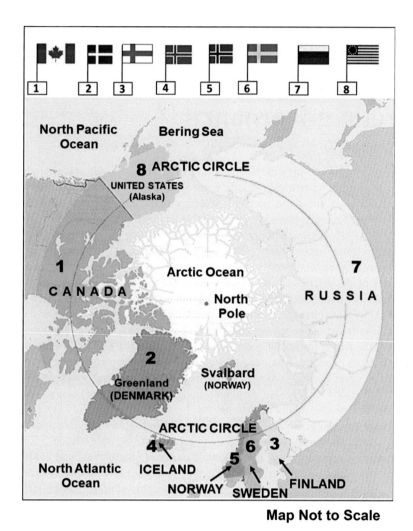

Map Not to Scale

FIGURE 11.1 Geographical location of all Arctic Nations surrounding Arctic Ocean. *Source: https://www.drishtiias.com/images/uploads/1568719428_image1.jpg.*

states should be based on respecting each other's rights, strengthening their communication, increasing mutual understanding, trust, support, and assistance for each other and seeking areas of converging interests.

Addressing transregional issues through joint research endeavors represents a major field of cooperation between the Arctic and non-Arctic states. Although some of the most critical Arctic issues are national, many issues are regional or transregional and relate to the environmental impacts of climate change, shipping, and resource development. These issues require a more comprehensive understanding of the causes and impacts of natural variability and human-induced environmental changes in the Arctic. The areas of international Arctic cooperation are endlessly increasing, making huge potential moreover as important challenges. Arctic cooperation began with attention on environmental protection and research, however quickly enlarged to embrace property development. Cooperation between the Arctic and non-Arctic states has continued to develop on a variety of levels, either bilaterally or inside the prevailing frameworks of regional forums and international organizations, on research, environmental protection, and property development. In accordance with the United Nations Convention on the Law of the Sea (UNCLOS) and alternative relevant international legal frameworks, Arctic states have sovereign rights and jurisdiction within their Arctic region, whereas non-Arctic states additionally fancy rights of research and navigation.

Arctic states, with a larger stake in Arctic-related issues, argue that they should play a more important role in Arctic affairs than non-Arctic countries. In the meantime, given the transregional implications of certain Arctic issues, non-Arctic states that fall under such influence argue that they have legitimate interests in Arctic-related issues. With their interests intertwined, there is no doubt that both Arctic and non-Arctic states will play increasingly significant roles in Arctic affairs. There is a risk that factors like the gap of recent shipping routes and the development of natural resources might become a cause for friction among states. It is vital to strengthen military presence within the region to combat resulting in tension and confrontations.

1. Spitsbergen

Spitsbergen (Coordinates: 78 degrees 45′N; 16 degrees 00′E) is the largest and only permanently populated island of the Svalbard archipelago in northern Norway (Fig. 11.2). Constituting the westernmost part of the archipelago, it borders the Arctic Ocean, the Norwegian Sea, and the Greenland Sea. Spitsbergen covers an area of 37,673 km^2, making it the largest island in Norway and the 36th largest in the world. The administrative center is Longyearbyen. Spitsbergen was covered in 21,977 km^2 of ice in 1999, which was approximately 58.5% of the island's total area. The island was first used as a whaling base in the 17th and 18th centuries, after which it was abandoned. Coal mining started at the end of the 19th century, and several permanent communities were established. The Svalbard Treaty of 1920 recognized Norwegian sovereignty and established Svalbard as a free economic zone and a demilitarized zone.

2. The Spitsbergen Treaty

Spitsbergen was no man's land until the early 20th century. Since the 1600s people from several nations have carried out various activities in Svalbard: hunting and trapping, research, mining, and tourism. For a long time, these

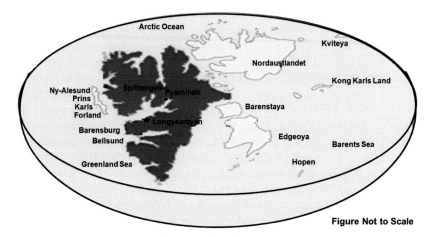

FIGURE 11.2 Svalbard archipelago with Spitsbergen shown in red. *Source: https://en.wikipedia. org/wiki/Spitsbergen#/media/File:Spitsbergen.png.*

activities took place without the region belonging to any state—Svalbard was a kind of international common land. This meant that there were no laws or regulations, and there were no courts to settle disputes. This functioned well if the activities consisted of hunting, trapping, fishing, and research. The region was large, and the conflicts were few. At the beginning of the 20th century, it was mainly the mining industry that created the need for changes. It was important to have sole rights to land and mineral deposits, and there arose a need for legislation and courts to settle disputes, for example between the mining companies and their workers.

Things changed in the late 19th and early 20th century when mining became the dominating field of economy in Spitsbergen. During the peace conferences in Paris amid the First World War, the Norwegians could convince other nations to put Spitsbergen under Norwegian sovereignty. This was formally done with the Spitsbergen Treaty, which was signed on February 09, 1920 in Versailles (Fig. 11.3). Contrary to its historical perspectives, the treaty is commonly referred to as "Svalbard Treaty."

In 1925, the treaty came into force and was included in Norwegian law. The "Svalbard law" (this is where the term "Svalbard" came in) came into force on August 14, 1925. This day is considered Svalbard's national day in Longyearbyen. But Spitsbergen did not become a part of Norway just as any other part of the mainland. The treaty defines several frame conditions, such as

- Spitsbergen is under Norwegian administration and legislation
- Citizens of all signatory nations have free access and the right of economic activities
- Spitsbergen remains demilitarized No nation, including Norway, can permanently station military personnel or equipment on Spitsbergen.

FIGURE 11.3 Fredrik Wedel Jarlsberg, Norwegian ambassador is signing the Spitsbergen Treaty on February 09, 1920 in Versailles, Paris. *Source: https://www.spitsbergen-svalbard.com/ spitsbergen-information/history/the-spitsbergentreaty.html.*

The treaty has served its purpose altogether well. It is the only one of all the treaties signed in 1920 in Versailles that is still in force and it is not substantially questioned by anyone, although there are critical voices concerning the issue.

This treaty had a relation with the Svalbard archipelago situated in the North of Norway. Therefore, it became popular as the Svalbard Treaty. According to this agreement, many countries got permission to establish centers for scientific research in the Svalbard archipelago.

In the beginning, the countries that were part of the treaty were Denmark, France, Italy, Japan, the Netherland, Norway, Sweden, the United States, and Great Britain (with all dominions on the other side of the sea). At that time, Australia, Canada, and India were included in the dominions of Britain. So, their representatives had also signed this treaty. Later, Russia and Germany also joined in 1924 and 1925, respectively. Now, the total members of this treaty were 40. According to this treaty, the archipelago of Svalbard has been considered as part of Norway. Now, Norway Government frames all the rules and regulations for it, but at the time of framing and implementing, the terms and conditions of the treaty are also considered. The citizens of all the signing countries possess all rights to enter and live in the Svalbard archipelago equal to the citizens of Norway. They also have an equal right to fishing, hunting, navigation-related works, mining, or for trading and scientific studies and research works. Although these activities should be under the law of Norway but in these matters no privilege is given to the citizens of any country.

According to the Governor of Svalbard "if one gets job in Svalbard, then according to the Svalbard treaty he/she automatically get right to migrate there."

According to the treaty, the responsibility of preservation of the natural environment of Svalbard is also handed over to Norway. Norway Government has formed Svalbard Environment Act for it which has been implemented from July 1, 2002, for the protection of the land of Svalbard, forest, wealth, animals and to make provision for sewage and disposal of waste material. The satisfactory implementation of the transport system of Svalbard is also included in it, but in that Act, there is full freedom to complete the research and occupational works without harming the environment.

The Svalbard Treaty is one of the few parts of the agreements reached Versailles that still have practical significance. Norwegian policies concerning Svalbard have always had an aim to respect the Treaty and to ensure that it is complied with to bring peace and stability to the area. The Treaty has, without a doubt, helped ensure that this has been a success. At present, around 2600 people live in Svalbard. These come from many nations, but the largest groupings are Norwegians living in Longyearbyen, Svea, and Ny-Ålesund and Russians and Ukrainians in Barentsburg.

The Treaty regulates research but states that such regulations shall be set forth under a separate agreement. Such an agreement has not yet been worked out. In practice—and in the spirit of the Svalbard Treaty—researchers from all countries are given equal rights to conduct research. During the last few years, considerable international research activities have developed on Svalbard.

As of January 2005, the following nations are registered as signatories to the Svalbard Treaty: Afghanistan, Albania, Argentina, Australia, Austria, Belgium, Bulgaria, Canada, Chile, China, Denmark, Egypt, Estonia, Finland, France, Great Britain, Germany, Greece, Holland, Hungary, Iceland, India, Italy, Japan, Monaco, New Zealand, Norway, Poland, Portugal, Romania, Russia, Saudi-Arabia, Spain, Switzerland, Sweden, South-Africa, The Dominican Republic, the United States, and Venezuela (Fig. 11.4).

3. Arctic Council

The Arctic Council is an intergovernmental forum for the eight Arctic states: Canada, the Kingdom of Denmark, Finland, Iceland, Norway, the Russian Federation, Sweden, and the United States, which are all currently jostling for ownership of the region's frozen seas. These states sought to promote the Arctic as a zone of cooperation and joint problem-solving while recognizing the distinct geography, demographics, and economics of the Far North. The Council's headquarters are in Tromsø, Norway. Established in 1996, the Arctic Council plays an advisory role by promoting cooperation, coordination, and

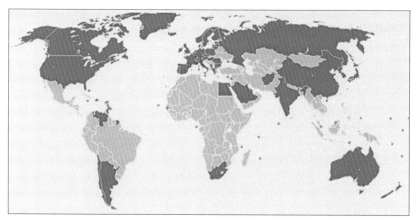

FIGURE 11.4 Signatory countries to the Spitsbergen Treaty. *Source: https://en.wikipedia.org/wiki/Svalbard_Treaty#/media/File:Svalbard_signatories.svg.*

interaction among the Arctic States. The Council is made up of member and observer states, Indigenous "permanent participants" and observer organizations [1].

In 1996, the Arctic Council an intergovernmental forum to "provide a means for promoting cooperation, coordination, and interaction among the Arctic States" and indigenous communities was established by the Ottawa Declaration. In addition to the five Arctic border countries, Finland, Sweden, and Iceland are full members of the Arctic Council. Members of the indigenous Arctic communities also hold permanent seats since part of the organization's mission is to support the sustainable development of the environment and maintain the health and well-being of its native peoples. As permanent participants, they have full consultation rights regarding the Council's negotiating decisions.

Twelve countries are permanent Observers, including six new countries as of May 2013, namely China, India, South Korea, Italy, Japan, and Singapore. The addition of these new members marked the first expansion of non-Arctic nations to Observer status since the Arctic Council was formed in 1996. Table 11.1 shows the status as of May 2019, about the Arctic Member States and the Observer States in the Arctic Council (Fig. 11.5). Whereas, Fig. 11.6 exhibits Status of the Composition, functionaries, and participants of the Arctic Council (as of May 2019).

However, at present, the major stakeholders in the opening of the Arctic region are the eight Arctic countries, all of whom are members of the Arctic Council, Canada, Denmark, Finland, Iceland, Norway, Russia, Sweden, and the United States, as well as the Council's 13 Observer States. These countries are actively exploring the geopolitical, strategic, and economic potential of the Arctic region. Switzerland was granted Observer status in 2017 [6].

TABLE 11.1 Status of the Arctic Member States and Observer States in the Arctic Council as of May 2019.

Sl. no.	Type of states	
	Member States	Observer States
	Only states with territory in the Arctic can be members of the Council [2]. Countries with sovereignty over the lands within the Arctic Circle constitute the members of the council	Observer status is open to non-Arctic states approved by the Council at the Ministerial Meetings that occur once every 2 years. Observers have no voting rights in the Council [3–5]
1	Canada	Germany, 1998
2	Denmark; representing • Greenland • Faroe Islands	Netherlands, 1998
3	Finland	Poland, 1998
4	Iceland	The United Kingdom, 1998
5	Norway	France, 2000
6	Russia	Spain, 2006
7	Sweden	China, 2013
8	United States	India, 2013
9	–	Italy, 2013
10	–	Japan, 2013
11	–	South Korea, 2013
12	–	Singapore, 2013
13	–	Switzerland, 2017

While the Arctic Council sees this expansion as an indication of the organization's increasing prestige and resources, new Observer countries were motivated by their growing economic and research interests in the region. Additionally, intergovernmental and interparliamentary organizations, as well as nongovernmental organizations (NGOs), have been inducted as Observers.

The Arctic Council has made significant research contributions to understanding the Arctic environment through Working Groups, long-term impact assessments, and specific task forces such as the Task Force for Action on Black Carbon and Methane. The Arctic Council has even established binding

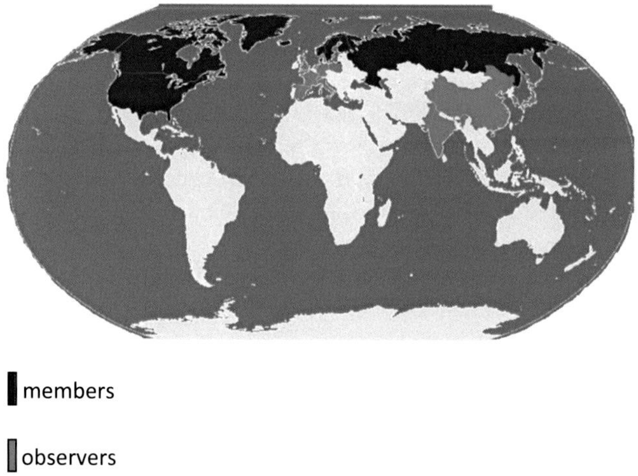

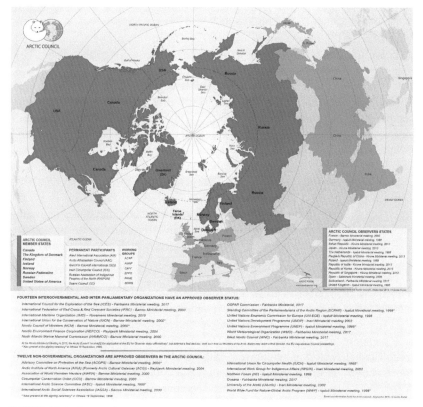

members

observers

FIGURE 11.5 The status as of May 2019, regarding the Arctic Member States and the Observer States in the Arctic Council. *Source: https://en.wikipedia.org/wiki/Arctic_Council#/media/File: ArcticCouncil.svg.*

FIGURE 11.6 Status of the Composition, functionaries, and participants of the Arctic Council (as of May 2019). *Source: https://arcticportal.org/images/news/2019/Arctic_Council_members_ and_observers_2019.jpg.*

multilateral agreements on regional issues such as search-and-rescue protocols and oil-spill response, etc. However, the Arctic Council is unable to enforce these treaties or coordinate practice exercises between signatory states due to its funding constraints and a self-imposed restriction against addressing "hard security." While security discussions on military presence and cooperation in the region have occurred bilaterally, they do not occur multilaterally as part of the Council. This illustrates the limits of the Council's reach in exercising Arctic governance. Further work is needed at the country level and through cross-country cooperation to develop response capabilities in the event of an oil spill or need for a rescue operation in the Arctic.

Beyond the Arctic Council's work on issues related to environmental research, search and rescue, and oil-spill response, the organization has recently moved to facilitate greater engagement with the private sector. In the year 2013, the Council established the Arctic Economic Council (AEC), under Canadian chairmanship, aimed at fostering business development in the region, engaging in deeper circumpolar cooperation, and providing a business perspective to the work of the Arctic Council. While some critics fear that the AEC will elevate business interests over those of environmental groups and indigenous peoples, many leaders see the potential for a business role in sustainable development in the Arctic.

In May 2013, Japan gained observer status in the Arctic Council (AC). Japan will further strengthen its contributions to the work of the council, for example, by dispatching experts and government officials to Working Groups, Task Forces, and other Council meetings. It will also examine further contributions that can be made through policy dialogues with the AC chair, member states, and others. From the standpoint of enabling further contributions to the AC, Japan will pay close attention to the discussions of the AC and how the role of observers is being examined within the AC. It will participate actively in discussions of expanding the role of Observers [7].

Development of the Arctic Council Participation in the Arctic Council is a central aspect of US Arctic policy. The Council is the key place in which diplomacy related to the region takes place. It has been that since its establishment in 1996, but in the period since then it has grown in importance tremendously. Its core focus remains on environmental protection and sustainability [8].

The United States Arctic Policy has a face in the form of the Arctic Council. Arctic Council is an authoritative international body on Arctic issues, and the United States has started to develop its own strategy toward addressing its resources and claims toward the region. The international aims of the policy are to support the ratification of the UNCLOS, promote participation in the Arctic Council, develop agreements with other Arctic countries on increased human activity in the region, and "continue to cooperate with other countries on Arctic issues through the United Nations (UN) and its specialized agencies." The policy argues against the need for a treaty among Arctic nations similar to the Antarctic Treaty restricting commercial and military activities. The policy calls for the United States to, "assert a more active and influential national presence to protect its Arctic interests and project sea power throughout the region," and to secure free passage of vessels through the Northwest Passage and the Northern Sea Route [9].

The Arctic Council is a forum of sort dedicated to Arctic issues. Each of the Arctic nation-states is represented viz. Canada, the United States, Denmark (Greenland), Russia, Iceland, Sweden, Norway, and Finland. Also represented are six Indigenous associations, called Permanent Participants, representing Inuit, Gwich'in, Athabaskans, Sami, Aleut, and 40 Russian indigenous groups' interests. The Model Arctic Council is a recent initiative to bolster the ranks of Arctic professionals with younger talent like students. The model that the Arctic Council works on is widely celebrated both for its emphasis on inclusion, which is not seen in many other places, and for how collaboratively the work is conducted. There is real wisdom in the structure of the Arctic Council, both real and model. Bringing together people from disparate backgrounds to participate openly in regional-level work is revolutionary and is perhaps the reason that the Arctic feels like such a sane zone of governance, despite increasing global turmoil and climatic challenges. The idea behind the Model Arctic Council is a really sound one that will no doubt bear fruit [10].

4. Arctic Council politics shift

Five of the eight nations in the group are also NATO members, whose charter commits member states to mutual military assistance. This may be the reason for the security issues not easily discussed by Arctic Council nation. To complicate matters, the Arctic Council is formally prohibited from discussing military security in the Arctic. Members consequently discuss security issues in informal meetings. Any discussion on addressing security concerns in the Arctic region. May not be easy unless such inhibitions are removed from the Arctic Council.

5. Asian interests

The developments on the security front are being perceived as an indication that the future militarization of the Arctic and such a situation will be uncomfortable for stakeholders other than the Arctic Council States who intend to stake control over the region's resources and sea lanes. Having realized the resource potential of the region and the significance of the Northwest Passage and the Northern Sea Route in the future as trade and energy transits, observer states perceive that in all matters strictly strategic, it is Arctic Council states that intend to control these matters.

Since they were granted "observer" status in May 2013, India, Japan, Singapore, and South Korea have become more conscious of the politics of the Arctic in the backdrop of the slow militarization of the region especially extremely conspicuous presence of China.

6. The Polar Code

The recently established "International Code for Ships Operating in Polar Waters" (Polar Code) also has a critical role to play in Arctic governance. The code was drafted by the International Maritime Organization (IMO), a specialized agency of the United Nations that concentrates on maritime regulation and got into effect in 2017 after final approval from IMO governing bodies. It covers a wide range of environmental and shipping safety measures, from ship design to search-and-rescue procedures. Because the code covers both the environment and shipping issues, it will be binding under both the International Convention for the Safety of Life at Sea and the International Convention for the Prevention of Pollution from Ships. All the Arctic nations are set to comply with the code, a robust signal that they acknowledge the importance of cooperation in shipping safety to all or any parties in operation within the Arctic region.

The Polar Code is a landmark agreement crafted over years of negotiation. It requires certification of ships sailing in polar waters, bans the discharge of oil and oily mixtures, and limits the release of sewage. However, the Polar Code does not ban heavy fuel oil in the Arctic, even though this type of fuel is already proscribed in the Antarctic. This issue is particularly salient because there are currently no methods to clean heavy fuel oil spills under Arctic conditions. The code also does not address black carbon emissions, invasive species, underwater noise, or preservation of seabird colonies.

7. Growing issues

While these governing bodies and international agreements play a vital role in Arctic governance, it is individual countries that may still have the foremost power within the region. Whereas cooperative governance, environmental protection, and security play a key role in every strategy, aspects associated with economic development are additionally outstanding than they were in the past. Property development remains the main target.

Finally, key stakeholders in the region are not limited to governments but personal companies, NGOs, environmental activists, and autochthonic populations all play a major role in Arctic relations. Greenpeace, a well-known NGO, has been terribly vocal in its objections to Arctic drilling.

Business and government leaders, notably those from non-Arctic countries, could succeed to establish the Arctic Circle in 2013, a forum to extend their participation within the debates encompassing the region as Arctic Council according to some provides disproportionate advantage to its eight permanent members. Since its origination, the Arctic Circle has served principally as a forum for business interests and a few supposed think tanks however have not advanced to come out with any policies or recommendations.

8. Outlook

The ways public and private stakeholders in the Arctic interact over the next 5 years will be critical to the future development of the region. As the United States' National Strategy for the Arctic Region strongly identifies the Arctic as a region that is "peaceful, stable, and free of conflict." To achieve this goal, continued cooperation and coordination between all stakeholders, countries, private corporations, indigenous peoples, and NGOs will be essential. However, for the Arctic to become a stable business operating environment, much work is still needed. Competing territorial claims, combined with changing regulatory standards, have created a difficult and uncertain operating environment, particularly for the industry. No doubt that the Arctic region presents enormous economic opportunities, but serious challenges remain alive.

It is noteworthy that the Arctic today is not only in the sphere of geopolitical interests of circumpolar Nations (Russia, the United States, Canada, Denmark, and Norway), but other States located far from the region (China, Japan, South Korea, India), as well as several international organizations that have not previously been involved in Arctic Affairs (NATO and EU). For Russia, it began the development of the polar region eight centuries ago, the Arctic remains today an integral part of the Russian geopolitical space [11].

9. Geopolitical dimensions of Arctic

Presently apart from traditional Arctic states, far more international organizations and non-Arctic states are showing an increased interest in the Arctic. The exclusivity of Arctic governance has been challenged by the activities of states from outside the region, such as the United Kingdom, France, Germany, China, Japan, South Korea, and India. These states are taking a special interest in many aspects of the Arctic that focus on scientific research, shipping, and resource development. Estimated oil and gas reserves in the continental shelves of the northern seas and visions of new trans-Arctic sea routes are also gaining the attention of transnational corporations that are increasingly interested in the potential commercial value of Arctic energy resources.

To gain an influence in the Arctic region, China implements a strategy that has worked well in Africa and Latin America, that is, investing and joining with local companies and financing good works to earn goodwill. Its scientists have become pillars of multinational Arctic research, and their icebreaker has been used in joint expeditions. It is a wrong way to think that the Arctic will only be of concern to those people living in the Arctic region. It is a concern to every nation, because there is no country that will escape the consequences, either through rising sea levels or extreme weather patterns. Therefore, there is a need for a common and comprehensive policy toward the challenges that are raised by ice melting in the North Pole. However, the activities and rhetoric of players who have expressed their interests in the Arctic region indicate the contrary. Obviously, the Arctic region has become a mirror of great powers' geopolitical interests [12].

When China along with Japan, South Korea, Singapore, India, and Italy was granted permanent Observer status in the Arctic Council, it left many experts wondering whether a paradigm shift in geopolitics is taking place in the region. Until recently, security issues, search and rescue protocols, indigenous rights, climate change, and other environmental priorities were the main concerns of the intergovernmental forums [13].

The Arctic has grown steadily as an important area for active diplomacy for the United States over the past years. Receding ice caused by climate change will bring greater human activity in the region, including increased shipping and energy exploration. The region is also an essential location for conducting the science that is needed to understand climate change. There are numerous topics relevant to the United States' interests in the Arctic, including economic and resource development, conservation of flora and fauna, and the rights of indigenous and local communities. Strengthen international cooperation. The strategy also lists four guiding principles viz. safeguard peace and stability; make decisions using the best available information; pursue innovative arrangements; and consult and coordinate with Alaska Natives. Under "Strengthen International Cooperation," the strategy provides for pursuing arrangements that promote shared Arctic State prosperity, protecting the Arctic environment and enhancing security; working through the Arctic Council to advance the United States interests in the Arctic region; acceding to the Law of the Sea Convention; and cooperating with other interested parties [14].

The geopolitical landscape of the Arctic today is a significant departure from the great power politics of the Cold War. Apart from traditional Arctic states, far more international organizations and non-Arctic states are showing an increased interest in the Arctic. It also reflects the longstanding expectations of researchers, countries, and international organizations involved in Arctic governance. China's policy goals in the Arctic are shaped in four key words—to understand, protect, develop, and participate in the governance of the Arctic [15].

10. Seed Vault

Norwegian Government has constructed a unique structure in Svalbard. It is a large "Seed Vault" (Fig. 11.7). The seeds of all crops being planted all over the

FIGURE 11.7 Svalbard global Seed Vault. *Source: https://upload.wikimedia.org/wikipedia/commons/9/97/Entrance_to_the_Seed_Vault_%28cropped%29.jpg.*

world and seeds of other botanical plants have been collected and stored in it. The main aim behind the construction of this structure is that if for any reason, or due to any disaster like accident, mismanagement, war global warming, etc. the biodiversity of all crops of the world ends, then it can be started again. By the cutting of rocks of Svalbard, a 120 miles long tunnel has been made for this Vault in which the temperature remains at 5°C naturally. Refrigerators are provided to keep the temperature of the tunnel up to 18°C. Seeds of various crops have been transported from 1400 godowns of various countries and they are stored. In view of safety, such doorways have been fixed in the tunnel that cannot be destroyed by the explosion. Altogether there is the provision of air-locks in it. India also donated seeds of Indian rice and wheat variety in 2008.

References

[1] About the Arctic Council. The Arctic Council. April 7, 2011. Archived from the Original on September 27, 2013. Retrieved Sep 6, 2013.

[2] Member States.

[3] Category: Observers (2011-04-27). "Six Non-arctic Countries Have Been Admitted as Observers to the Arctic Council". Arctic-council.org. Retrieved 2013-09-24.

[4] India Enters Arctic Council as Observer. The.hindu.com. 2013-05-15. Retrieved 2013-09-24.

[5] Ghattas, K. (2013-05-14). "Arctic Council: John Kerry Steps into Arctic Diplomacy". Bbc.co.uk. Retrieved 2013-09-24.

[6] D. Nanda, India's Arctic Potential", ORF Occasional Paper No. 186, February 2019), 2019.

[7] Japan's Arctic Policy by the Headquarters for Ocean Policy, the Government of Japan on 16 October 2015, and Edited by Embassy of Japan in Finland.

[8] E.T. Bloom, United States perspectives on the arctic, in: D.A. Berry, N. Bowles, H. Jones (Eds.), Governing the North American Arctic. St Antony's Series, Palgrave Macmillan, London, 2016. https://link.springer.com/chapter/10.1057/97811374939101.

[9] https://en.wikipedia.org/wiki/Arctic _policy_of_the_United_States#cite_note-10.

[10] https://jsis.washington.edu/news/indigenous-influence-arctic-council-learning-role-playing/ Indigenous influence on the Arctic Council: Learning through role playing, September 1, 2017, Michael Brown.

[11] Russia's Geopolitical Interests in the Arctic. Print Version Material Posted: Publication Date: 15-11-2014.

[12] V. Mārtiņš, 15 Augusts, http://www.lai.lv/viedokli/arctic-a-mirror-of-great-powers-geo political-interests-305, 2013.

[13] https://www.theguardian.com/environment/2013/jun/04/china-arctics-mineral-riches.

[14] E.T. Bloom, United States perspectives on the arctic, in: D.A. Berry, N. Bowles, H. Jones (Eds.), Governing the North American Arctic. St Antony's Series, Palgrave Macmillan, London, 2016. https://link.springer.com/chapter/10.1057/9781137493910_12.

[15] ICAS Report, Report: China's Interests in the Arctic: Opportunities and Challenges, March 16, 2018. By Nong Hong, https://chinaus-icas.org/report/chinas-interests-in-the-arctic-opportunities-challenges/.

Chapter 12

International research initiatives of Arctic

Active participation in response to global issues regarding the Arctic and the formulation process of international rules for the Arctic has emerged amid growing concern over the impact of environmental changes in the Arctic on the environment of the Earth as a whole, including global warming and climate change, countries actively convey the findings of its scientific observations and research, and work toward examining the possibility of enabling a new agenda based on wide-ranging international cooperation. This chapter will cover details on such international scientific endeavors by various participating countries having scientific interest in the Arctic region. Before providing an insight into International scientific efforts in the Arctic region, it is appropriate to understand the Svalbard and Ny-Ålesund areas of the Arctic where such scientific cooperation is being nurtured since long and Indian scientific activities are presently focused.

1. Svalbard

Svalbard is a group of islands (archipelago) located between the Arctic Ocean (Fig. 12.1), Barents Sea, Greenland Sea, and the Norwegian Sea. Since 1920, these islands are an integrated part of Norway. These islands are located directly north of Norway in the Arctic Ocean. The group of islands ranges from 74 to 81 degrees north latitude and from 10 to 35 degrees east longitude. The islands can be divided into two groups: the Spitsbergen group of Barentsøya, Edgeøya, Nordaustlandet, and Prins Karls Forland, and the more remote islands of Bjørnøya, Hopen, Kong Karls Land, and Kvitøya.

Svalbard is the northernmost place in the world with a permanent population. Located between the 76 and 81 degrees parallels, they are far more northerly than any part of Alaska and all but a few of Canada's Arctic islands. In fact, they would be permanently locked in by ice if not for the moderating influence of the Gulf Stream, and it is this comparative warmth that makes them habitable. The islands cover a total of 62,050 km^2, the largest of which are Spitsbergen, Nordaustlandet, and Edgeøya. The combined permanent population is less than 3000, nearly all of which is concentrated in the main settlements of Longyearbyen and Barentsburg on Spitsbergen.

The Arctic. https://doi.org/10.1016/B978-0-12-823735-9.00004-7

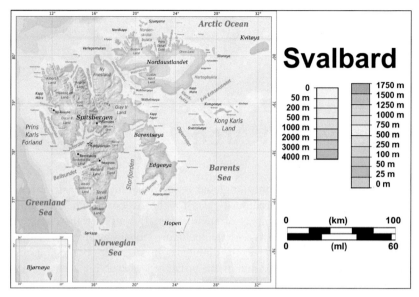

FIGURE 12.1 Map of Svalbard archipelago. *Source: https://en.wikipedia.org/wiki/Woodfjorden.*

The archipelago features an Arctic climate, although with significantly higher temperatures than other areas at the same latitude. The flora takes advantage of the long period of midnight sun to compensate for the polar night. Svalbard is a breeding ground for many seabirds, and features polar bears, reindeer, the arctic fox, and certain marine mammals. Seven national parks and 23 nature reserves cover two-thirds of the archipelago, protecting the largely untouched, yet fragile, natural environment. Approximately 60% of the archipelago is covered with glaciers, and the islands feature many mountains and fjords.

1.1 Ny-Ålesund

Ny-Ålesund situated at 78 degrees 55′ N, 11 degrees 56′ E on the west coast of Spitsbergen in the Svalbard archipelago is a functional base for international research and environmental monitoring settlements (Fig. 12.2). Ny-Ålesund was a mining town for commercial exploitation of the coal deposits in the Kongsfjorden area from 1916 to the closure of the mines in 1963. A former coal-mining colony, Ny-Ålesund is about 2100 km north of Oslo. After the mining activity, Ny-Ålesund developed into a research village providing laboratories, facilities, and research infrastructures to scientists from many nations interested in polar research. During the summer period, the population grows to over 180 when scientists from around 20 nations will arrive to carry out their research. Eleven institutions from ten different countries have permanent research stations in Ny-Ålesund. Out of which three research stations are

FIGURE 12.2 A panoramic view of Ny-Alesund. *Source: https://en.wikipedia.org/wiki/Ny-%C3%85lesund https://en.wikipedia.org/wiki/Ny-%C3%85lesund.*

continuousely operational on year round basis. The distance to the North Pole is 1231 km. The mean temperature in the coldest month (February) is −14°C, while the warmest month (July) has a mean temperature of +5°C. Ny-Ålesund on the west coast is an international center for various modern Arctic research activities. The village is one of the world's northernmost human settlements and is surrounded by glaciers, moraines, rivers, mountains, and a typical tundra system. Most of the fauna living in Svalbard is represented in the area with birds (auks, kittiwakes, terns, barnacle geese, reindeer, foxes, polar bears, and often visible in the fjord seals and sometimes belugas and walrus.

There is no permanent population living and science is the only currency in the Ny-Alesund region. The local community is mostly a mixture of staff from the Kings Bay Company and scientists of various nationalities. Visits of tourists coming mostly onboard cruise ships for just a few hours are relatively frequent in summer. The closest town is Longyearbyen (population c. 2000). During the summer season, the Ny-Ålesund population reaches 150−180 persons down to the 30−40 permanent staff during the winter months. A regular air shuttle service organized by Kings Bay AS with a small plane (14 passengers) connects Longyearbyen with Ny-Ålesund. The flight takes approximately 25−30 min. Access is also possible by ship but there is no regular ship transport to Ny-Ålesund. Transport of freight is possible with a monthly freight ship excepted during winter. Fjord shores and islands are easily accessible using small boats and local transportation is possible by cars, snowmobiles, or bicycles.

To maintain communication and interaction among many countries and their research institutions, a Ny-Alesand Science Management Committee was established in 1994. The Committee comprises the representatives of all permanent and facility providing institutes. This committee provides consultation to Kings' Bay AS research program. Kings' Bay AS is an organization of Norwegian Government. It is the owner of Ny-Alesand and administers it. It is responsible for providing all public facilities to Ny-Alesand, encouraging research and scientific programs and developing Ny-Alesand as International

Arctic Research Center. As it is night during the whole winter, the residents spend most of their time inside their houses. During that time, they read, attend gyms, and see video films. During summers, the population starts increasing and it reaches 150.

Ny Island is not a family station. There is no school, market, or hospital available there. There are no medical or doctor facilities. If anybody falls sick, the Governor's helicopter is provided to take the patient to a hospital in Longyearbyen city situated 107 km away. There is no question about caste, creed, nationality, and religion there. All people take food together. The food is nutritious keeping in view the environmental conditions and facilities available there. Vegetable food is also provided on some occasions. This region is historically important and has got the honor of housing northernmost post-office (Fig. 12.3). The old postoffice in Ny-Ålesund is not in function anymore. Today, the Kongsfjordbutikken serves as a postoffice, where you can get stamps and send mail from the postbox on the outside wall. The stamps in the old postoffice are not official stamps; they are just souvenir stamps.

The Town and Mine Museum (Fig. 12.4) at Ny-Ålesund is housed in a building constructed in 1917, located in Ny-Ålesund. The museum is, originally for storage and later got converted into the town store in 1920. The museum was created long afterward, in 1988. Things featured in the remodeled museum, which is unstaffed for the most part, include exhibits on the history of Ny-Ålesund since the site was first visited in the 1600s, the coal mining industry, the 1962 accident, early aviation on Svalbard, the scientific research carried out in Ny-Ålesund, and the settlement's society and culture.

FIGURE 12.3 Kongsfjordbutikken "The symbolic Post Office" at Ny-Alesund. *Source: https://upload.wikimedia.org/wikipedia/commons/e/ed/Ny-Alesund_post_office_1.JPG.*

FIGURE 12.4   Town and mine museum at Ny-Ålesund. *Source: https://upload.wikimedia.org/ wikipedia/commons/8/84/NOR-2016-Svalbard-Ny-%C3%85lesund-Museum_01.jpg.*

2. International Research Center

Now, the Norwegian Government has developed the Nye-Alesand as the base of Natural Science Research and Arctic Research. In view of this, all the necessary auxiliary facilities have been provided there. In the decade of 1990, many research institutions were established there out of which some are from Norway and some from other countries. Approximately scientists from 20 countries are conducting research in them. Kings Bay AS owns and manages Ny-Ålesund research village. Eleven institutions from ten countries around the world have established research facilities here, three of which are permanently manned. In addition, several more institutions and nations come regularly to Ny-Ålesund to carry out research field activities. Table 12.1 provides the details of various International research station being operated by different countries in the Ny-Alesand area [1].

For the study and research of various factors of the Arctic region, the Indian research center "Himadri" is also situated there. Details of Some main research centers of Ny-Alesand are provided below.

2.1 Norwegian Polar Institute

The Norwegian Polar Institute (NPI) occupies the Sverdrup (Fig. 12.5). The current building was built in 1999 but NPI has been carrying out research in Ny-Ålesund since 1968. NPI offers its services to all Norwegian institutions. It also hosts projects from countries that do not have their own facility in Ny-Ålesund. Sverdrup maintains instruments for long- and short-term monitoring programs, offers logistical services, and research facilities to visiting scientists.

TABLE 12.1 Details of various operational research stations located in Ny-Alesund [1].

Station	Institution	Nationality	Est.	Research
Amundsen–Nobile	National Research Council of Italy	Italy	2009	Atmospheric
Arctic	University of Groningen	Netherlands	1995	Ecology and others
Arctic Yellow River	Chinese Arctic and Antarctic Administration	China	2004	Environment, glaciology, meteorology, marine ecosystems, meteorology, space –Earth measurements
British	British Antarctic Survey	United Kingdom	1991	Earth and life sciences
Corbel	Polar Institute Paul Emile Victor	France		Atmospheric sciences
Dasan	Korea Polar Research Institute	South Korea	2002	Atmospheric chemistry glacial and periglacial geomorphology and hydrology
Dirigibile Italia	National Research Council of Italy	Italy	1997	Environment and climate
Himadri	National Centre for Antarctic and Ocean Research	India	2008	Atmospheric sciences, marine ecosystems and pollution
Japanese	National Institute of Polar Research	Japan	1990	Atmospheric physics, glaciology, meteorology, oceanography and terrestrial biology
Koldewey	Alfred Wegener Institute	Germany	1991	Atmospheric physics, biology, chemistry and geology. Part of the Total Carbon Column Observing Network.

TABLE 12.1 Details of various operational research stations located in Ny-Alesund [1].—cont'd

Station	Institution	Nationality	Est.	Research
Marine Laboratory	Kings Bay	Norway	2005	Marine biology
Rabot	Polar Institute Paul Emile Victor	France	1999	Atmospheric and life sciences
Rocket Range	Andøya Rocket Range	Norway	1997	Sounding rockets
Sverdrup	Norwegian Polar Institute	Norway	1999	Various
VLBI	Norwegian Mapping Authority	Norway	1992	Very-long-baseline interferometry
Zeppelin	Norwegian Polar Institute	Norway	1988	Atmospheric

FIGURE 12.5 Sverdrup Norway research station. *Source: https://upload.wikimedia.org/wikipedia/commons/8/86/Sverdrupstasjonen_1.JPG.*

This center provides different types of facilities to many national and international research centers. It also governs the Environmental Chemical Monitoring Center situated on the 470-m height from sea level on Zeppelin Mountain. NPI owns the building of the Zeppelin Research Observatory on the Zeppelin Mountain, a couple of kilometers south of the settlement. The observatory is located on Zeppelin Mountain, close to Ny-Ålesund in Svalbard. At 79 degrees N, the station lies in an undisturbed arctic environment, far from

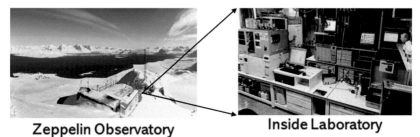

Zeppelin Observatory   Inside Laboratory

FIGURE 12.6 Norwegian Zeppelin Observatory at Ny-Alesund. The unique location of this station makes it an ideal platform for the monitoring of global atmospheric change and long-range pollution transport. *Source: https://commons.wikimedia.org/wiki/File:ZeppelinLandscape.jpg.*

major sources of pollution. The observatory's unique location makes it an ideal platform for monitoring changes in the global atmosphere and pollution. Norwegian Institute for Air Research (NILU) is responsible for the scientific activities at the Zeppelin Observatory (Fig. 12.6).

It continuously provides information to different universities about static magnetic fields, polar luminescence, glacier science, earth quack monitoring, and environment. It also provides the logistic facilities for the field researchers. The permanent engineers, technicians of this institute also provide the facility of different special instruments to the different scientific projects and institutes the whole year.

The following are measured at Zeppelin.

- Greenhouse gases (CO, CO_2, CH_4, and N_2O)
- Chlorinated and fluorinated greenhouse gases (more than 30), which are also covered by the Montreal Protocol (ozone-depleting compounds)
- New greenhouse gases such as some HFCs and SO_2F_2
- Hydrocarbons
- Aerosols (physical and optical properties) in close collaboration with Stockholm University
- Ozone layer and UV radiation
- Inorganic components (e.g., sulfur and nitrogen compounds) and ground-level ozone
- Organic matter and trace elements
- Organic environmental pollutants (PAH, PCBs, HCB, DDTs, HCHs, OCPs, BDEs)
- Mercury and trace elements
- NILU

NILU measures the concentration of carbon dioxide in the atmosphere with the help of the Atmospheric Science Centre of Stockholm University. This Institute also conducts the serials of Air compression in the center located at a Joplin mountain.

2.2 Norwegian mapping authority

The Norwegian Mapping Authority has been permanently based in Ny-Ålesund since 1992 and operates the VLBI-antennae close to the airstrip. The antenna is a part of a global network of VLBI-antennas. Results from the measurements inform about the Earth's rotation speed, helps in the definition of boundaries, and opens for the prediction of earthquakes and tsunami. This Institute also provides the facilities of geomathematics, geodetic research. This institution has close ties with other organizations and shares scientific information with them.

2.3 The Svalbard Rocket Range or SvalRak

It is a launch site (Fig. 12.7) for sounding rockets at Ny-Ålesund in Svalbard, Norway. The site is suitable for dayside Polar Cusp, Cleft, and Camp for rocket launching. The first rocket was launched in 1997 using this site and is owned by Andøya Space Center, which is owned by the Norwegian Ministry of Trade, Industry and Fisheries, and the Kongsberg Group.

SvalRak's location at the 79th parallel north makes it well suited for launching rockets to investigate the Earth's magnetic field. It is used mostly by American, Japanese, and Norwegian researchers. It is the world's northernmost launch site. In addition, it is suitable for studies of dayside Aurora and magnetic region activities.

2.4 Alfred Wagner Institute and Institute Polaris Francis

The German Alfred-Wegener-Institut für Polar-und Meeresforschung (AWI) has been operating Koldewey station (Fig. 12.8) at Ny-Alesund since 1991. AWI focuses on biology, chemistry, geology, and atmospheric physics and can provide bedrooms, office space, and a living room at the base. In addition,

FIGURE 12.7 The Svalbard Rocket Range SvalRak used for launching sounding rockets. *Source: https://en.wikipedia.org/wiki/Svalbard_Rocket_Range.*

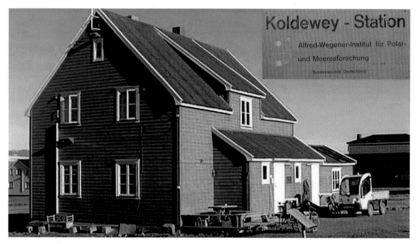

FIGURE 12.8 German Research Station "Koldewey at Ny-Alesuds." *Source: https://www.awi. de/en/expedition/stations/awipev-arctic-research-base.html.*

AWI operates the NDACC Observatory that is used for studies of physics and chemistry of the troposphere and stratosphere.

A specially designed lab was set up in the center in 1994. In 1999, a French Robot Centre was established by Institute Polaris Frances, an Institute of France. The responsibility of this center was to conduct scientific programs in Polar and sub-Polar Regions. In 2003, the German and French institutes made a joint center where scientists from both countries could work together. This joint center carries out networking about the changes in the stratosphere during the whole year. Committed to long-term planning, this center is a member of international networks including World Meteorological Science Organisation.

2.5 German-French Research Station AWIPEV at Ny-Alesund

The French Institut Paul Émile Victor (IPEV) established the Charles Rabot (Rabot) in 1999. IPEV has been mostly interested in atmospheric and life sciences since the start. IPEV also runs the Jean Corbel (Corbel) located about 5 km, south-east of Ny-Ålesund. Corbel runs almost entirely on green energy, and it is often described as the "clean". In 2003, AWI and IPEV merged their operations (logistics and administration) in Ny-Ålesund and created a joint station named AWIPEV (Fig. 12.9) and has a permanent staff of three persons. Buildings are owned and provided in Ny-Ålesund by the Norwegian company Kings Bay AS. This base includes the Koldewey Station, the Rabot Station, and the Jean Corbel Station buildings.

The station provides logistical resources: boats, snowmobiles, vehicles, storage facilities (cold and warm, cooling facilities), field equipment, workshop

FIGURE 12.9   German-French Research Station AWIPEV at Ny-Alesund. *Source: https://www. awipev.eu/.*

and office facilities and scientific resources: laboratories, scientific equipment, and sampling tools. The research activities cover a large field of scientific disciplines. Both, long-term and short-term projects at AWIPEV, which includes atmospheric studies, marine and terrestrial biology, or cryosphere studies.

Besides Station AWIPEV also undertake different measurements as well as field expeditions such as atmospheric long-term measurements, scientific diving, glacier expeditions, seabirds as indicators of global changes in the marine ecosystems, installation of measuring systems in the permafrost and in the fjord, and launches of research balloons [2].

2.6 Japanese Research Station at Ny-Ålesund

The Arctic Environment Research Center (AERC) was established in June 1990 at the National Institute of Polar Research (NIPR) to promote the study of sea ice, oceanography, marine ecology, terrestrial ecology, atmospheric sciences, glaciology, and upper atmospheric sciences. Since April 2015, AERC has started enhancements of international research planning and cooperative works. NIPR operates a research station at Ny-Alesund (Fig. 12.10) on Spitsbergen Island, Svalbard Islands (79 degrees N, 12 degrees E), with support from the Norwegian Polar Institute (NPI) since 1991. The station has engaged in the observations of clouds, aerosols, radiation, aurora, greenhouse gases, vegetation distributions, and ecosystem studies within the international cooperative observation system in Ny-Ålesund.

FIGURE 12.10   Japanese Research Station at Ny-Ålesund. *Source: https://www.nipr.ac.jp/ english/arctic/center.html.*

2.7 Natural Environment Research Council

UK Arctic Research Station (Fig. 12.11) at Ny-Ålesund, is situated at Lat. 78 degrees 55'0″N and Long. 11 degrees 55'59″E and has been operational since 1991. This institute studies the changes in the hydrosphere and territorial sciences worldwide. This station is funded by Natural Environment Research Council (NERC) and managed and operated by British Antarctic Survey, and is situated in the international research community at Ny-Ålesund on the high Arctic island of Spitsbergen, part of the Svalbard archipelago.

The station has been involved in several EU infrastructure projects and is currently a partner in the EU-funded project INTERACT (International Network for Terrestrial Research and Monitoring in the Arctic). INTERACT is a network of 83 terrestrial field bases in northern Europe, Russia, the United States, Canada, Iceland, Greenland, the Faroe Islands, and Scotland as well as stations in northern alpine areas. As such, the Station is available to non-UK-based researchers wishing to apply to research Ny-Ålesund via the INTERACT Transnational Access scheme.

FIGURE 12.11   UK Arctic Research Station at Ny-Alesund. *Source: https://www.arctic.ac.uk/uk-arctic-research-station/.*

2.8 Italian Research Station Dirigibile Italia at Ny-Ålesund

The National Research Council of Italy (CNR) established their base in Ny-Ålesund named Dirigibile Italia (Fig. 12.12) in 1997. The Italian Arctic Station Dirigibile Italia is a multidisciplinary research facility located in Ny-Ålesund in the Norwegian Archipelago of Spitzbergen (Svalbard). The station is managed by the CNR and the activities are coordinated by the Polar Support Unit of the CNR Department of Earth and Environment: POLARNET. Dirigibile Italia supports the CNR's studies on climatic changes. Research focuses much on environmental and climatic studies of ice and marine sediments, and research on interaction mechanisms among atmosphere, biosphere, hydrosphere, and geosphere. The Dirigibile Italia is not permanently manned.

2.9 Dasan (South Korean Research Station)

An agreement was signed by Norwegian Government with South Korea to set up a Research Center in Ny Alesand. The Korean Polar Research Institute (KOPRI) opened their research Station Dasan (Fig. 12.13) in Ny-Ålesund in 2002. Research activities focus mainly on environmental research, glacial and periglacial geomorphology, hydrology, and atmospheric chemistry. Dasan is not permanently manned.

2.10 Chinese Research Center "Yellow River" at Ny-Alesund

Chinese Arctic and Antarctic Administration inaugurated the Yellow River (Fig. 12.14) in 2004. Since then activities have been increasing within a wide range of science with research on meteorology, space-Earth measurements, glaciology, marine ecosystems, and Arctic environment. Yellow River is not permanently manned but has visiting scientists during large parts of the year. The Polar Research Institute of China took over the administration and management of the Yellow River in 2017.

FIGURE 12.12 Italian Research Station Dirigibile Italia at Ny-Ålesund. *Source: https://www.cnr. it/en/peculiar-research-structures.*

FIGURE 12.13 South Korean Research Station Dasan at Ny-Alesund. *Source: http://www. arcticstation.nl/maps/building.php?nr=15.*

FIGURE 12.14 China's Research Station "Yellow River" at Ny-Alesund Station. *Source: http://www.arcticstation.nl/maps/building.php?nr=18.*

2.11 Arctic Center of UiG The Netherlands at Ny-Alesund

To promote multidisciplinary research in the Arctic region that covers a wide range of science but has also a strong focus on ecology, notably the ecology of the barnacle goose, the Arctic Center (Fig. 12.15) of the University of Groningen (UiG) in the Netherlands established a research station in Ny-Ålesund in 1995. The station is closed during the winter.

FIGURE 12.15   Arctic Center of UiG The Netherlands at Ny-Alesund. *Source: http://www. arcticstation.nl/maps/building.php?nr=47.*

3. Shared research infrastructure

Ny-Ålesund offers a wide range of shared scientific infrastructure that can easily be made available to International Scientists for instrumentation at any of these facilities. A brief account of such shared research facilities is provided below.

3.1 Kings Bay Marine Laboratory

The Marine Laboratory (Fig. 12.16) that is equipped for a range of experimental research with system control of experimental variables like room and water temperatures is developed as an International laboratory to conduct

FIGURE 12.16   Kings Bay Marine Laboratory. *Source: https://www.npolar.no/en/.*

research in the field of marine ecology, physiology, and biochemistry, and physical sciences like oceanography, marine geology, and ice physics. It is also appropriate for experiments under ambient conditions such as seawater and light. This facility is owned, maintained, and operated by Kings Bay AS.

3.2 Zeppelin Observatory

The Zeppelin Observatory (Fig. 12.17A and B) at 475 m above the sea level is set up for atmospheric research and monitoring as an integral part of a network of important global observatories for atmospheric measurements, and several regional and global monitoring networks. The Observatory provides several air inlets, and meteorological instruments, radiation instruments, and other samplers mounted on the roof. The building itself consists of several rooms, where the users keep their instruments including a room, the "campaign-room," exclusively reserved for project-based measurement campaigns. It is owned and managed by the NPI.

FIGURE 12.17A Zeppelin Observatory. *Source: https://blogrecherche.wp.imt.fr/en/2018/07/02/ reactive-trace-gases/the-zeppelin-observatory-in-ny-alesund-svalbard-arctic-view-from-the-rooftop-2/.*

FIGURE 12.17B The gondola that runs up Zeppelin. *Source: https://en.wikipedia.org/wiki/ Zeppelinfjellet#/media/File:Zeppelin_cable_car.jpg.*

3.3 Amundsen-Nobile Climate Change Tower

To investigate energy budget in the surface layer, PBL dynamics and exchange fluxes (heat, momentum, chemicals) at the atmosphere—land interface, the Climate Change Tower with a height of 32 m host has been set up at Ny-Alesund (Fig. 12.18). The tower platform compiles 17 modules and every third floor is connected to power and optical fiber. The fiber cable is connected to the CNR station to allow real-time control and monitoring of measurements.

3.4 Gruvebadet Atmosphere Laboratory

The Gruvebadet Aerosol Laboratory (Fig. 12.19) is located nearby the village of Ny-Alesund, in the Svalbard archipelago. It is equipped to host aerosol sampling for measurements of chemical, physical, and optical properties. It is managed by CNR-ISAC personnel and operative from March to October each year. Gruvebadet is a common atmospheric laboratory located 1 km outside Ny-Ålesund. It was renovated in 2013. The building has several inlets on the roof, and host now atmospheric and aerosol instrumentations from CNR and NCAOR.

FIGURE 12.18 Amundsen-Nobile Climate Change Tower. *Source: https://upload.wikimedia. org/wikipedia/commons/d/da/Amundsen-Nobile_Climate_Change_Tower.jpg.*

FIGURE 12.19 Gruvebadet Atmosphere Laboratory. *Source:http://nysmac.npolar.no/research/research-infrastructure.html.*

3.5 Light sensitive cabin

The light sensitive cabin has recently been established in Ny-Ålesund (Fig. 12.20). The observatory, called jovially "lysbua" by Norwegians, accommodates four domes on the roof, two of them with slightly larger working space in the cabin than the two others. The domes were mounted on the roof and the cabin was placed at the designated site about 2.5 km away from a town having minimal light contamination. Two domes are already in use. One hosts the all-sky-camera (moved from the Rabot station) owned by Italy. The other

FIGURE 12.20 Light sensitive cabin. *Source: http://nysmac.npolar.no/research/research-infrastructure.html.*

one is occupied by the KOPRI with the Fabry-Perot Interferometer installed. The cabin is connected to high-speed internet, and it is insulated to keep the temperature inside above freezing.

4. Indian Arctic Research Center—Himadri

India's first research station located at the International Arctic Research base, NyÅlesund, Svalbard (Spitzbergen), Norway. It is located at 1200 km from the North Pole. It is operational since July 1, 2008. This establishment of the Indian Research Station "Himadri" has been toward sustenance and furtherance of scientific initiatives of Indian scientists in pursuit of Arctic scientific studies. Station "Himadri" is well-equipped laboratory that also supports various fieldworks required for pursuing research activities in the Arctic. It is located at the old school building in the Norwegian town of Ny-Ålesund, midway between Novaya Zemlya and Greenland, the manned research station is in one of the world's northernmost human settlements and represents an ideal land-based entrance to the European Arctic. Indian Arctic Research Station "Himadri" is adequately equipped with state of the art facilities that will be used for conducting round the year scientific research in contemporary fields of Arctic science with special emphasis on climate change. Being situated at 78 degrees 55′ N, 11 degrees 56′E, at Ny-Ålesund, Indian Research Station offers the ideal land-based entrance to the Arctic. Current areas of research in Ny-Ålesund include marine science, aurora physics, biology, glaciology, geology, environmental science, geodetic studies, rocket probe studies, atmospheric physics, terrestrial studies, and climate change monitoring, among others.

The Himadri base station (Fig. 12.21) has been managed by the National Centre for Polar and Ocean Research (NCPOR), Goa (India), an autonomous institute under the Ministry of Earth Sciences, Government of India. This premier Polar institute based in Goa has been coordinating the entire gamut of Antarctic research for the country.

With the opening of Himadri (the Indian research base), India has become the 10th country to have established its full-fledged research station at Ny-Ålesund, other countries to have similar research facilities in the region are already discussed above. The environmental research and monitoring programs presently running at the stations in Ny-Ålesund are concentrated mainly on atmospheric and biospheric research, with significant emphasis also laid on ionospheric research, solid earth physics, and glaciology, with permanent monitoring of the mass-balance of glaciers.

The station was set up in a refurbished two-floored building with four bedrooms. The building has an area of 220 m^2 (2400 sq ft) and has other facilities including a computer room, storeroom, drawing room, and internet. It can host eight scientists at normal conditions (Fig. 12.22). The crew of the station is given training in shooting with rifles to protect themselves from polar bears.

FIGURE 12.21 Indian Research Station "Himadri" situated at Ny-Ålesund, Svalbard. *Source: http://www.ncaor.gov.in/.*

FIGURE 12.22 A window to inside layout (A–F) of the Indian Research Station "Himadri" situated at Ny-Ålesund, Svalbard.

References

[1] http://www.ncaor.gov.in/.
[2] http://www.awipev.eu/awipev-observatories/.

Chapter 13

Indian Arctic science program

It is natural to be curious as to why our scientists are interested in a distant place like the Arctic? The Arctic is a region with extreme climate and India had no lingual, religious or cultural relations with its native as well. It is true that India had no commercial, economic, religious, or lingual bondage with Arctic people. It is, therefore, not these facts per se that are important for India's motivation to be in the Arctic. India's role so far in the region has been restricted to the scientific arena. India has established considerable experience in polar science and research. India is currently an observer in the Arctic Council and has asserted that unlike other countries, its interest in the Arctic region is only driven by scientific interests to enhance its scientific expertise, "particularly its polar research capabilities" [1]. Following more than four decades of experience in scientific research in Antarctica (Southern High latitude regions), India set up its first research station "Himadri" in Ny-Ålesund, Spitsbergen (Arctic) way back in 2008, primarily to conduct scientific investigations on "glaciology, atmospheric sciences and biological sciences." Since many studies have pointed toward the connection between the Indian monsoon and the Arctic atmospheric processes, the importance of monitoring the effects of climate change in the Arctic has been acknowledged by many Indian scientists [2,3]. Scientific cooperation and science diplomacy are expected to continue for India's engagement with the Arctic region in the future as well.

Undoubtedly, there are other reasons for that. The Arctic Ocean and its nearby regions not only affect the climate of the whole earth, but it has preserved the ancient history of climate. They also indicate that what will be the future impact of the current activities by humans. The thermohaline circulation that starts from the Southern Part of the Arctic Ocean regulates the sea streams apart from regulating the climate of the earth. Therefore, the changes in this circulation affect the climate of Earth, sea level, biodiversity, and other factors. Though the Arctic region is located far from India, yet it is important for the subcontinent of India. Various studies indicate that there is a distant relationship between the North Polar Region and monsoons. Though this distant relationship is still a puzzle for the scientists and to solve the puzzle there is a need for detailed analytical studies. Indian scientists are eager to conduct such studies. This eagerness also increases because India was also a part of the

The Arctic. https://doi.org/10.1016/B978-0-12-823735-9.00001-1

189

Svalbard Treaty. Though it had not utilized the opportunity to conduct research in Svalbard, but now it is ready to do so.

Indubitably, there is considerable similarity in the natural climates of the Arctic and Antarctic. Seamlessly, the Indian scientists, who have gained extensive expertise and experience in the contemporary fields of polar science through India's rich heritage of 40 odd year's presence in the Antarctic arena, are ably and aptly utilizing now in the Arctic region. The commencement of the Arctic Research expedition was commenced in November 2006 when the then Minister of Science and Technology and Earth Sciences traveled to Norway. After the visits of the representatives of both the countries and mutual discussions, a concrete plan was prepared. The responsibility to implement the plan was given to the National Center for Polar and Ocean Research (NCPOR), Goa, India. Nevertheless, the actual implementation of the Indian Arctic Research Program was carried out in 2007 during International Polar Year. The International Polar Year was organized by Scientists, Unions of International Council and World Meteorological Organization in 2007–08. The Polar Year was also the 25th Anniversary of the International Geophysics Year (1957–58). It will be worth mentioning here that during the International Geophysics year, various scientific explorations and scientific studies were carried out at the world level with international cooperation, and the Antarctic Treaty was formed in 1959. The objective of the Polar year was to establish a relationship between the Polar region with the rest of the world. Strengthening of the relationship among climatic change, natural calamities, environmental protection, and sustainable human development and other global environment and human concerns were on the priory advocating the cross-cutting approach.

Another important incident also occurred during the International Polar Year 2007–08 relating to international cooperation. It was Science Pub Multinational Ocean Exploration Journey. The aim of the journey was to find out the major changes in the Arctic Ocean during the past years to predict future happenings. Dr. Neloy Khare, one of the authors of this book, also participated in this journey. This journey was conducted in Norway's expedition ship "RV. Lance" in August 2008. The sample of the sea level was taken during Science Pub. Based on this sample, studies were conducted about the ancient environment of the Arctic.

In the first phase of the Arctic Research Program, in August/September 2007, a five-member team of Indian Scientists visited Ny-Alesund's International Arctic Research Center and took account of the facilities available there. After the analysis of the facilities available in the Arctic Research Center, the scientists started various research projects in the field of Environmental Science, Microbiology, Geology, Glacier Science, and other associated fields toward data collection.

After the positive responses of the studies undertaken during the summer of 2007, the Indian scientists visited Ny-Alesund and started various research projects in the field of Astronomical Physics, Environmental Sciences, Biology, Microbiology, and Ancient Environment Sciences, etc.

1. India's scientific interest in the Arctic

India launched its first scientific expedition to the Arctic Ocean in 2007 and opened a research base named "Himadri" (Fig. 13.1) at the International Arctic Research Base at Ny-Alesund, Svalbard, Norway in July 2008 for carrying out studies in disciplines like Glaciology, Atmospheric sciences, and Biological sciences. The major objectives of the Indian Research in the Arctic Region are as follows:

○ To study the hypothesized teleconnection between the Arctic climate and the Indian monsoon by analyzing the sediment and ice core records from the Arctic glaciers and the Arctic Ocean.
○ To characterize sea ice in the Arctic using satellite data to estimate the effect of global warming in the northern polar region.
○ To conduct research on the dynamics and mass budget of Arctic glaciers focusing on the effect of glaciers on sea-level change.

FIGURE 13.1 (A–F) Different views of the Indian Research Station "Himadri" in Arctic. *Source: http://www.ncaor.gov.in/.*

o To carry out a comprehensive assessment of the flora and fauna of the Arctic and their response to anthropogenic activities. In addition, it is proposed to undertake a comparative study of the life forms from both the Polar Regions.

India has been closely following the developments in the Arctic region in light of the new opportunities and challenges emerging for the international community due to global warming-induced melting of the Arctic's ice cap. India's interests in the Arctic region are largely scientific and environmental issues, as well as strategic. Toward further strengthening the Indian perspectives of Arctic science, India has also entered Memorandum of Understanding (MoU) with the Norwegian Polar Research Institute of Norway, for cooperation in science, and with Kings Bay AS (A Norwegian Government-owned company) at Ny-Alesund for the logistic and infrastructure facilities for undertaking Arctic research and maintaining Indian Research base "Himadri" at Arctic region.

In 2019, India has been reelected as an Observer to the Council. However, at present, India does not have an official Arctic policy, nevertheless, its Arctic research objectives have been focused on ecological and environmental aspects, to better understand climate change occurring over the Arctic region and its impact on the global climate system with special reference to the tropical region including India.

1.1 Indian gateway to Arctic research "Himadri"

The Indian scientific endeavors in the Arctic realm commenced when a five-member scientific team visited Ny-Ålesund on the Svalbard archipelago of Norway during the summer of 2007. India leased a station building at Ny-Ålesund from Kings Bay AS that owns and manages the facilities at the International Research Base. As stated earlier, the Indian station "Himadri" (Fig. 13.1) was inaugurated on July 1, 2008 by the then Hon'ble Minister of Science and Technology and Earth Sciences, in the presence of dignitaries from Norway, the United Kingdom, Germany, and other countries besides India.

Considering the immense scope for scientific research, NCPOR, as the nodal agency for the Indian Arctic Program entered a MoU with the Norwegian Polar Institute for scientific cooperation in Polar Sciences. After the successful achievements of the first year in Arctic science, India's proposal to become a member of the Ny-Ålesund Science Managers Committee (NySMAC) was accepted in November 2008. The major role of NySMAC is to enhance cooperation and coordination among research activities at the Ny-Ålesund International Arctic Research and Monitoring Facility. The-then Hon'ble Minister of Science and Technology & Earth Sciences, together with the then Hon'ble Minister of Research and Higher Education of Norway, led a high-level delegation to Ny-Ålesund on June 6, 2010. At Himadri, both the Ministers formally launched the Indian Arctic web portal. Major developments of the Indian Arctic program were laid in the year 2013, when India received the permanent observer status in the Arctic Council during the meeting of the Arctic Council in Kiruna, Sweden with the efforts of the then Foreign Minister [4].

As mentioned earlier, various Indian Scientists wanted to conduct research in various fields in the Arctic Region. In this context, they have expressed their interest to the concerned authorities. Considering the interest of the scientists and the facilities available at the International Arctic Research Center at Ny-Alesund, took steps to take a building on a lease, where the Indian scientists can set up a laboratory to conduct research. The total land of Ny-Alesaud is under the control of Kings Bay AS. Therefore, a building of 220 sq. meter was taken on 5-year lease from Kings Bay AS in January 2008. Various modifications were made in the building. The first floor was converted in Laboratory and the first floor was used for the accommodation of the scientist. In this way, India became the 10th country to establish its research center in the Arctic.

In addition to Norway, Sweden, France, Germany, Netherland, Italy, Korea, and China have already established their research centers in Ny-Alesund. The then Science and Technology & Earth Sciences Minister also visited "Himadri" to motivate the Indian Scientist in June 2010.

1.2 Cooperative agreement

Himadri is managed by Goa-based NCPOR. To encourage the research and scientific studies in the Arctic region, NCPOR has entered into a Cooperative Agreement with the Norwegian Polar Institute. The agreement is related to the research of the following:

1. Geological mapping and related studies
2. Biochemistry of sea ice system
3. Environmental Physics and Chemistry
4. Glacier studies
5. Ancient Climatic studies

India wants to implement its scientific programs as multiinstitutional experiments by utilizing available facilities at the International Arctic Center at Ny-Alesund. Initially, the studies will be focused on the areas near the Ny-Alesund, and later the Arctic Ocean and Svalbard Islands will be included in it. Every year, during spring and summers, the Indian scientist groups will visit the Arctic to collect information. (It is very difficult to stay in Ny-Alesund during winters; therefore, the scientists work during spring and summers only). Indian scientists will ensure that within the framework of the Agreement with the Norwegian Government, they develop the geographical area of their studies.

It is hoped that Indian scientists will include other subjects in their program to collaborate with the other countries in their scientific programs. Simultaneously, it is proposed to establish scientific equipment like Ionosonde Brand, Band Seismograph, etc. It will enable to collect samples to conduct experiments for Geochemical and Hydrochemical studies. Scientists from other countries can also avail these facilities. It will be important to mention here that the plan developed by the Indian Scientists is based on their 1-year experience in Ny-Alesund and three decades experience in the Antarctic (1981—2010). It also has been ensured that there is no duplication of efforts by other scientists working in Ny-Alesund.

Indian researchers have been exploring the Arctic from metering the precipitation in the Arctic region to drawing up a baseline data on microbial biodiversity in Kongsfjorden sediments and carrying out a biochemical evaluation and biomarker characterization of Arctic fjord sediments, Indian researchers are into a series of scientific investigations in the icy terrains of Arctic.

2. Arctic science—Indian contribution

Working from "Himadri," country's Arctic research station at Spitsbergen, Norway, the Indian scientists are looking into the various aspects of life that thrives there to gain a better understanding about the climatological factors that influence the Arctic weather and its impact on climate changes elsewhere.

The Indian researchers are looking into the variability of precipitation over the Arctic by measuring it using Micro Rain Radar. The temperature and humidity profiles of the region have been created using a microwave radiometer profiler.

It will help fill the gaps in the observational data on Arctic clouds, especially during the dark winter season, by recording measurements using a ceilometer to measure the vertical profiles of multiple cloud layers. The long-term monitoring of the Kongsfjorden system of the Arctic region for climate change studies has been the mandate of the Indian Science activities over the Arctic region.

Additionally, attention is being paid to the long-term environmental monitoring of fjord ecosystems, especially the ecosystem of Kongsfjorden and Krossfjorden. The water and sediment chemistry with respect to climate change in the fjords has been the focus area of research. The biochemical evaluation and biomarker characterization from Arctic fjord sediments have been made. The zooplankton ecology and planktonic food web dynamics in Kongsfjorden has been studied using in-situ and satellite oceanography techniques. The impact of glacial runoff and associated arctic freshening on microbial community structure has been assessed from the Kongsfjorden. While trying to understand the macrobenthic faunal composition at selected locations in the Kongsjorden Fjord covering a length of 15 km from the oceanic end to the glacial end of the Fjord, the impact of glacial run-off and associated Arctic freshening of the microbial community structure of Kongsfjorden has also been evaluated. Some salient findings of Indian scientific initiatives in Arctic research are provided below.

2.1 Production of carbon monoxide from ice packs

Carbon monoxide is the most important atmospheric gas, which is produced due to the combustion of fossil fuel. It is also produced in large amount by industries and motor vehicles. Carbon monoxide is a poisonous gas and has a very low composition in the atmosphere. This is because the hydroxyl radical (OH) gets combined with it chemically and converts it into nonpoisonous material. It helps in monitoring the quantity of hydroxyl compounds. Hydroxyl being an oxidizer controls the composition of many greenhouse gases in the atmosphere. Recent studies show that carbon monoxide is continuously produced and liberated in large amounts in glacier areas. Our scientists also conducted experiments related to carbon monoxide at Indian Research Center, Maitri, which is situated in Antarctic Islands. With the help of different experiments, they come to know about the regular production cycle of carbon monoxide because of regular consideration of Solar Actinic rays. Consequently, scientists came to know that the production of carbon monoxide is due to a photochemical reaction in Antarctic Glaciers.

It is considered that some organic materials like formaldehyde (HCHO), which is entrapped in ice crystals, get decomposed due to photochemical reactions and as a result the product formed acts like a lustrate for CO production. The measurement of Carbon Monoxide, were according to conventional information which prove that formaldehyde is present in large amount in glaciers. Our scientists want to compare different conclusions that were obtained from predefined strategy about the production of carbon monoxide from snowpack's in the Antarctic field. The main instruments on which experiments are generally conducted in Polar Regions are carbon monoxide, oxygen, nitrogen oxide analyzer, solar photometer, portable climate center, pyrometer, etc.

Our scientists regularly perform different experiments to produce carbon monoxide from snowpacks and come to know about the regular alterability in carbon monoxide production in Ny-Alesund (2008) especially in the month of March and August. Despite all, our scientists also measure the concentration of black carbon, composition of Aerosoles, their size, distribution, and the water vapor in the atmospheric air of Ny-Alesund. In the summer of 2008, an ozone analyzer has also been included for the measurement of surface ozone concentration.

2.2 Benthic studies

Benthic foraminifera are the established indicators of climate change and have been used extensively to infer immediate past climatic variations. Studies carried out in the Kongsfjorden system reported that the abundance of foraminifera increases from the inner-most fjord to the mouth in Kongsfjorden [5,6].

2.3 Microbial animals of Arctic

Microorganisms have been found to be useful especially in the field of Agriculture and Medicine. They can easily synthesize many biomolecules that were very difficult to be synthesized in chemical laboratories. Scientists always make efforts to get the information related to different properties and characteristics of microorganisms and they want to get familiar with the microorganism having special characteristics in different areas and conditions. The Arctic is the only region that comprise many glaciers, snowpacks, marine snow, swampy areas of Tundra, and different oceans. Certainly, the microorganisms of these regions must have some special characteristics. Some Indian scientists conducted experiments on different microorganisms of these regions to find out whether they can be useful for industries or not.

2.4 Microbes of sea ice

Sea ice is one of the most important components of Polar Regions that affects most of the animals and marine environment of the region. Sea ice surrounds approximately 5% part of the Northern Hemisphere and 8% of the Southern hemisphere. Out of the totally ice-covered areas of the earth, approximately 67% part is covered with sea ice but in terms of volume, the sea ice is only 1% of the total ice. A very thin layer of ice is deposited on the sea, which is easily affected by the temperature of the environment and sea activities. The Arctic is a deep basin and some parts of it are permanently covered with ice. Generally, the continental shelf and shallow marginal branches also get covered with sea ice during the winter season. Approximately 50%−70% part of the Arctic is covered by ice permanently. It is 3-m-deep in sea and 2 m in coastal areas.

On the contrary, 90% of ice of the Antarctic Ocean gets melted in the initial year and its thickness is less than 2 m. Though the qualities of the ice of both oceans are similar, they have some special features also. The microorganisms get affected by these special features. Generally, virus, bacteria, diatoms, fungi, and protozoa are some of the microorganisms that reside in sea ice. They spent their whole life in the frozen sea ice. Generally, among all microbes' studies are conducted mainly on parasite bacterial and single-cell algae.

Indian scientists are mainly conducting studies to know the following facts:

i. Ice packs of Arctic and Antarctic oceans and concentration, density of different colonies of microbes, their diversity, and way of survival.
ii. Surface of sea ice, its internal part, and different varieties of microbes and their way of survival, especially those that reside in basins of oceans.
iii. The main mechanism through which different animals get mingled in sea ice.
iv. Contribution of a different consortium of microbes of sea ice in biogeochemical reactions.

To fulfill the above objectives, Indian scientists are taking steps in search of an appropriate place near Ny-Alesund. They measure the depth and density of sea ice over there. The sea ice samples are taken and sent to India at 50°C. They were examined well with the help of the most prestigious classification method and recognized by different molecular techniques. As a result, we can also find out the different nutritional components as well as quantities of chlorophyll along with salinity.

2.5 Microbial community of Kongsfjorden system

The microbial community composition was investigated by using high-throughput 16S rRNA gene amplicon sequencing. Investigations have revealed that microbial community compositions are significantly different spatially within the inner and outer fjord of the systems [7,8].

2.6 Electrical properties of Ny-Alesund and atmosphere

The Indian scientists are continuously working and studying the electrical conductivity of atmosphere, small and medium ions quantities, and concentration, size, and their distribution. These studies are being conducted since the summer of 2007. These types of studies are generally more beneficial and important for the information related to the worldwide electrical path and solar static relationship. The main objective of measurement of aerosoles is to find out the sources of aerosoles, their concentration, and production of aerosoles in that region. These measurements will provide information about the basic pollutants and their arrival in Europe and other Northern regions of the Arctic.

2.7 Studies of glaciers

Many changes have been noticed in previous years of Glaciers of the Arctic. Indian scientists also take much interest to get the information related to these changes in glaciers. They also studied West Craig, Borger, and Mindralavan Glaciers in this reference.

2.7.1 Geomorphology and sedimentology of diversified morphological zones of glaciated terrain of the Ny-Alesund

The Ny-Alesund region is characterized by diversified surface processes that carve the landscape and so exhibits variable and complex landforms. Extensive work has been carried out on climate change using various proxies; however, no much attention has been paid in understanding the climate events using geomorphological and sedimentological parameters. In the present paper sediment characteristics, AMS 14C dates and geomorphic features have been used for the reconstruction of paleoclimate. On the basis of the distribution of landforms, and sediments, this region has been classified into five morphological zones such as glacial (moraines GL), proglacial (lacustrine deposits LD), outwash plain (sandur deposits OWP), fluvial deposits (FD), and coastal cliff (CC). The geomorphic analysis and sedimentary parameters revealed that GL consists of unconsolidated, unstratified, massive, devoid of any sedimentary structures, coarse-grained, matrix-supported boulders, whereas OWP, LD, FD, and CC are semiconsolidated, stratified, fine-grained, layers of sand, silt, and clay with gravels and faint sedimentary structures. The sediments of CC and LD are very poorly sorted, very positively skewed, very leptokurtic, medium to fine sand, silt, and clay. The sediment characteristics of various morphological zones suggest that this region was carved and dominated by glaciers and paraglacial processes. It might has occurred under cold climate at ice stages during 47.5, 38, 23, 18, 8.8, 6.1, 1 ka BP and under warm climate at interglacial stages during 44, 27, 12, 10.5 ka BP. The poorly sorted sediments for all the morphological zones explain the fluctuating energy of the depositional environment and so the prevailing climate was not consistent and persistent for a long period of time [9].

2.7.2 Quartz grain microtexture and magnetic susceptibility assessment of Ny-Alesund region

The quartz grain microtexture reveals predominant glacial activities in the top 40 cm of the section, while the middle 40−55 cm part represents some aeolian activities along with glacial signatures. The bottommost part, in addition to glacial markers, exhibits some aqueous evidence as well. The lithology shows medium-grained sand in the upper part and coarse-grained sand with occasional shell pieces in the lower part. Angular gravels (2−12 mm) are present throughout with increasing size from top to bottom. On the basis of the earlier observations and 14C AMS dates, it can be summarized that after the Last Glacial Maximum (LGM), the pre-Holocene period shows rapid glacial retreat, followed by a warmer period during the early Holocene. Mid- and late Holocene is marked by a predominantly glacial environment characterized by meltwater streams originating from the glaciers and flowing into the fjord. Magnetic susceptibility studies have also been attempted and four alternate stages of colder and warmer phases have been established. Though some similarities among the different climatic phases are discernible between the quartz grain microtexture and magnetic susceptibility studies, they are not completely compatible, which is probably due to their different responses to the climatic variations [10].

2.7.3 Remote sensing observations and model reanalysis

Applications of remote sensing techniques along with the modeling have been applied to quantify the magnitude of Arctic sea-ice loss in the boreal summer (July−September), especially in September at different timescales (daily, monthly, annual, and decadal). The investigation on the accelerated decline in the Arctic sea-ice was performed using different datasets of passive microwave satellite imagery and model reanalysis. Arctic sea-ice declined rapidly in the boreal summer (−10.2 0.8%decade1) during 1979−2018, while, the highest decline in sea-ice extent (SIE) (i.e., 82,300 km^2 year1/-12.8 1.1 %decade1) is reported in the month of September. Since late 1979, the SIE recorded the sixth-lowest decline during September 2018 (4.71 million km^2). Incidentally, the records of 12 lowest extents in the satellite era occurred in the last 12 years. The loss of SIE and sea-ice concentration (SIC) are attributed to the impacts of land-ocean warming and the northward heat advection into the Arctic Ocean over the past 40 years (1979−2018) while substantial warming rates have been identified in the Arctic Ocean in the last 40 years. The prevailing ocean-atmospheric warming in the Arctic, the SIE, SIC, and SIT have reduced, resulting in the decline of the sea-ice volume (SIV) at the rate of −3.0 0.2 (1000 km^3 decade1). Further, it is observed that the SIV in September 2018 was three times lower than in September 1979. This study demonstrates the linkages of sea-ice dynamics to ice drifting and accelerated melting due to persistent low pressure and high air-ocean temperatures, supplemented by the coupled ocean-atmospheric forcing [11]. Accelerated decline is recorded in the Arctic sea ice extent and sea ice concentration over the past four decades. The Ocean-atmosphere coupled mechanism plays an important role in the global climate change. Sea ice variability and trends were computed using satellite and model reanalysis measurements for the total Arctic and each of their nine regions: (1) Seas of Okhotsk and Japan, (2) Bering Sea, (3) Hudson Bay, (4) Baffin Bay/Labrador Sea, (5) Gulf of St. Lawrence, (6) Greenland Sea, (7) Kara and Barents Seas, (8) Arctic Ocean, and (9) Canadian Archipelago. Overall Arctic sea ice declined in all seasons and on a yearly average basis, although the highest and lowest negative trends were recorded in summer and winter/spring, respectively. The study reveals that regionally mainly four sectors, Arctic Ocean, Kara and Barents Seas, the Greenland Sea, and the Baffin Bay region are majorly responsible for the negative sea ice extent trend for the total Arctic as a whole. The study demonstrated the interannual and seasonal variabilities of Arctic sea ice and interactions among the atmosphere, ice, and ocean [12].

2.7.4 Assessment of spatiotemporal variability of snowmelt across Svalbard

Indian researchers have taken up monitoring of snowmelt over the Svalbard region as significant changes in the interannual variation of Arctic snow and sea ice are connected to changes in the global climate using Active microwave sensors. These sensors are frequently used to detect surface melting because of their sensitivity to the liquid water presence in snow/ice. The annual melt duration and summer melt onset for the Svalbard archipelago has been mapped using microwave scatterometers flown on QuikScat, OSCAT, ASCAT, and OSCAT-2, providing one of the longest and continuous records of radar backscatter to estimate snowmelt onset and melt duration on Svalbard spanning 2000–17. A single threshold-based model was used to detect the timing of the snowmelt; the threshold was calculated using meteorological data from the manned weather stations. The results capture the timing and extent of melt events caused by warm air temperature and precipitation, because of the influx of moist, mild air from the Norwegian and Barents seas. The highest melt duration and earlier melt onset occurred in southernmost and western Svalbard, in response to the influence of warm west Spitsbergen Current. Compared to earlier studies, we found considerable interannual variability and regional differences. Though the record is short, there is an indication of the increasing trend in total days of melt duration and earlier summer melt onset date possibly linked to the general warming trend (Fig. 13.2A–D). Climate indices such as Interdecadal Pacific Oscillation and Pacific Decadal Oscillation are well correlated with onset melt and duration across Svalbard. With the reported year-after-year decrease in sea ice cover over the Arctic Ocean, the trend toward longer snowmelt duration inferred from this study is expected to enhance the Arctic amplification [13].

2.7.5 Assessment of mass balance of the Arctic glaciers

The study [14] presents changes in the area from 1993 to 2018 and mass balance from 2011 to 17 of Vestre Broggerbreen glacier, Ny-Ålesund, Arctic (Fig. 13.3). The glaciated area has decreased from 3.96 km^2 (in 1993) to 3.57 km^2 (in 2018), at a varying rate, resulting in a total area loss of 0.39 km^2. A comparatively rapid decrease in the glaciated area was found during 1998–2010 whereas less retreat rate was found in 1993–98. Mass balance of Vestre Broggerbreen glacier was found to be negative throughout the entire study period (2011–18). Mass balance ranged between −0.08 m w.e. (2013–14) and −1.22 m w.e (2015–16) with a cumulative mass balance of −4.31 m w.e (0.016 km^2 a–1). A strong relationship between mass balance and summer temperature was found with R2 = 0.97 at $P < .05$ [14]. This study presents changes in the area from 1993–2018 and mass balance from 2011–17 of Vestre Broggerbreen glacier, Ny-Ålesund, Arctic. The glaciated area has decreased from 3.96 km^2 in 1993 to 3.57 km^2 in 2018, at a

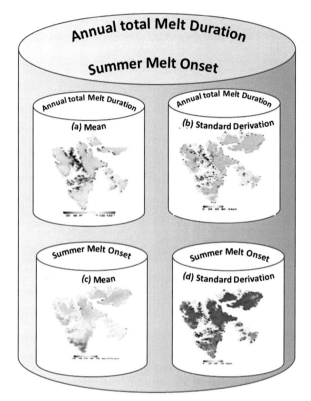

FIGURE 13.2 (A−D) The 18-year mean (2000−17) and standard deviation (SD) for melt duration (MD) and Svalbard melt Onset (SMO) over entire Svalbard region. *Source: A. J. Luis, personal communication.*

varying rate ranging between 0.011−0.02 km^2, resulting into a total area loss of 0.39 km^2 ($\sim$10% at 0.016 km^2 a−1). A comparatively rapid decrease in the glaciated area was found during 1998−2010 (0.02 km^2 a−1) whereas less retreat rate was found in 1993−98 (0.011 km^2 a−1) and 2010−18 (0.012 km^2 a−1). Mass balance of Vestre Broggerbreen glacier was found to be negative throughout the entire study period (2011−18). Mass balance ranged between −0.08 m w.e. (2013−14) to −1.22 m w.e (2015−16) with a cumulative mass balance of −4.31 m w.e (0.016 km^2 a−1). A strong relationship between mass balance and summer temperature was found with R2 = 0.97 at $P < 0.05$.

2.8 Studies on full solar eclipse

The total solar eclipse was noticed on August 1, 2008 in the Arctic Region. Indian scientists also analyzed and studied regarding the full solar eclipse.

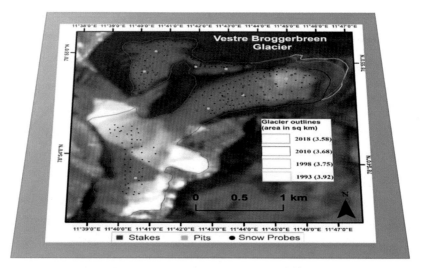

FIGURE 13.3 Location map of Vestre Broggerbreen glacier, ablation stakes, accumulation pits 67 and snow probes, Ny-Ålesund, Svalbard, Arctic. *Source: A.L. Ramanathan, personal communication.*

2.9 Studies on Kongsfjorden

Kongsfjorden, an icy archipelago having a length of about 40 km and width ranging from 5 to 10 km, is a glacial fjord in the Arctic (Svalbard). It lies on the northwest coast of Spitsbergen, the main island of Svalbard, and is a site where warmer waters of the Atlantic meet the colder waters of the Arctic. Being an open fjord without sill, it is largely influenced by the processes on the adjacent shelf. The Transformed Atlantic Water (TWA) from the west Spitsbergen current and the glacier-melt freshwater at the inner fjord creates strong temperature and salinity gradients along the length of the fjord. Southerly winds will produce down-welling at the coast and cause hindrance on exchange processes between the shelf and the fjord, while the northerly winds will move the TWA water below the upper layer toward the coast. The meltwater during summer not only stratifies the upper water column but significantly alters the turbidity.

This would have a profound influence on the seasonality in the phytoplankton biomass and primary production. Thus, an altered interaction between the Atlantic water with the (turbid) meltwaters from tidal glaciers on a seasonal to interannual time-scale is likely to affect the pelagic ecosystem in the fjord. Alternately, the benthic ecosystem is more likely to be affected by long-term changes in the fjord hydrography and sedimentation.

Against the above backdrop of the climate sensitivity of the fjord system, NCAOR has initiated an ambitious multi-institutional program of long-term monitoring of the Kongsfjorden by deploying an Ocean-Atmosphere mooring system with regular repeat transects to measure physical and biogeochemical parameters on a seasonal scale. The overall objective is to establish a long-term comprehensive physical, chemical, biological, and atmospheric measurement program to study.

The variability in the Arctic/Atlantic climate signal by understanding the interaction between the freshwater from the glacial run-off and Atlantic water from the west Spitsbergen current. The effect of interaction between the warm Atlantic water and the cold glacial-melt fresh water on the biological productivity and phytoplankton species composition and diversity within the fjord.

The winter convection and its role in biogeochemical cycling. The trigger mechanism of spring bloom and its temporal variability and biomass production, and the production and export of organic carbon in the fjord with a view to quantify the CO_2 flux.

To achieve the above objectives, a two-prong measurement strategy needs to be adopted.

To collect long-term time-series data on oceanographic (currents, temperature, salinity, turbidity) and biological parameters (PAR, O_2, fluorescence) through deploying an ocean mooring. Repeat transects to monitor the variability in the physical and biogeochemical parameters on an intraseasonal to interannual time scale. The in situ measurements during repeat transect monitoring are being carried out to cover three seasonal transitions.

As a prelude to the initiation of the above project, systematic CTD casts and surface sediment sampling were carried out during the summer months of May–July 2010 from different locations extending from the mouth to the head of the fjord. Repeat measurements were continued at these predefined stations through the three seasons of 2011 and 2012 [15].

2.9.1 Marine biogeochemical studies of Kongsfjorden

Study of the interactions of glacier meltwater and oceanic influence and their associated physical, biological, and geochemical processes is being carried out at Kongsfjorden located at Ny-Ålesund as an established reference site for Arctic marine studies, considering the uniqueness of the fjord, its proximity to the Arctic and the Atlantic water masses. This program is multi-institutional and multidisciplinary in nature that addresses the possible response of this fjord to climate variabilities at different time scales, like seasonal, annual, and long-term changes in its hydrography, hydrochemistry, and associated aspects.

2.10 Arctic pollution monitoring studies

As a part of Indo-Arctic scientific expeditions, an Indian study reveals about 180% rise in Aerosol Optical Depth (the parameter which gives the measure of aerosols like urban haze, smoke particles, desert dust, sea salt, etc.) over a period of 10 years with a major contribution from north-western Canada. It has been found that the anthropogenic pollution levels were the highest in a decade over the arctic region during the summer of 2011. The measurements were collected over Indian Arctic research station, "Himadri." These findings revealed "Unusual high values of aerosol optical depth evidenced in the Arctic during the summer of 2011 [16,17]." The temperatures in the arctic region are known to be rising at twice the rate, critically altering both human and natural systems. Such unprecedented warming due to pollution over the Arctic region results in retreating or cracking sea ice, it is restructuring Arctic ecosystems and permitting new industrial access for commercial fishing, off-shore energy, and commercial shipping on an alarmingly increased scale further adding up to pollution over the region.

The reason behind the rise in pollution is long-range transportation from the north-western Canadian side. However, why it was so only during 2011 remained an enigma. Domination of fine-mode particles corresponding to man-made pollutants was found compared to coarse particles of sea salt which ideally dominate the Arctic's atmosphere. Additionally, the results show consistent low-alpha values which revealed that the atmosphere was loaded with pollutants. The source of pollutants reaching the arctic was known to occur by long-range transportation via the European region. However, during the summer of 2011, the major contribution was from north-western Canada. The results of their experiment were confirmed by the measurements from the Aerosol Robotic NETwork (AERONET), operated by NASA and worldwide national scientific agencies at Hornsund, about 232 km away from Ny-Alesund, that registered a similar rise during the period.

Sedimentological and geochemical studies from the sediments of Arctic lakes and fjords (Krossfjord and Kongsfjord). Lakes and fjords are essential components of high latitude regions in which sediments received from different geological and glacial units exhibit almost similar characteristics as there is an intermixing of the sediments of various units due to ice, melt-water, and wind activities. Therefore, a multiproxy approach was adopted to identify the source of sediments and understand the depositional processes, occurring in lakes and fjord as these depositional environments act as a sink for nutrients and metals. For the study carried out, sediment samples collected from the Arctic and Antarctic region were analyzed for grain size, total organic carbon, total nitrogen, total phosphorus, biogenic silica, calcium carbonate, clay minerals, bulk, and speciation of metals along with the estimation of Beryllium concentration. Surface sediment samples from the Krossfjord and Kongsfjord, Arctic showed that coarse-grain fraction is higher in Kongsfjord as compared to the Krossfjord as it is largely influenced by different tidewater glaciers debouching in the fjord. Sediment core from the mouth of the Kongsfjord showed a high concentration of silt suggesting deposition of finer sediments through suspension mode. Sediment cores collected from the lakes of the Arctic region consist of >50% of sand in the upper portion of all the four cores (LA, L-1, L-2, L-3) which suggested the dominance of mechanical (glacial) weathering/frost weathering processes releasing coarse-grained material to the lake basin in the recent years. Organic components of the surface sediment samples from the Arctic region showed low nutrients concentrations in the inner fjord as it is in the proximity of the glacier and glacial water is devoid of nutrients and supplies a huge amount of coarser material. Both the fjords showed higher C: N values in shallower regions because of the presence of the high amount of terrestrial material and their association with coarse-grained sediment suggested grain size to be a dominant factor regulating the distribution of organic matter. A sediment core from the mouth of the Krossfjord showed a fluctuating trend of organic matter with depth indicating a changing rate of supply of organic matter through changing processes. Sediment cores from the lakes of the Arctic showed that TOC and TN were higher than the average value in the upper portion of core L-1 and L-2 suggesting high productivity due to the exposure of the lakes to the ice meltwater influx. C/N ratio varied from 15.50 to 38.32 suggesting a mixed (autochthonous and allochthonous terrestrial) source of organic matter. In the fjords, major and trace metals showed a decreasing trend on moving away from the coast like that of sand indicating their similar terrigenous source except for Cd and Ni. Speciation study showed a high concentration of Mn and Co in the labile phases of sediments, which can detrimentally affect sediment-associated biota. Therefore, there is a need to monitor the changes in the concentration of metals and protect the pristine environment. Sediment cores from lakes of the Arctic region (Ny-Alesund and Kuadehuken) showed that the elemental concentrations have increased toward the surface suggested either increased meltwater influx in recent years or diagenetic remobilization. Thus, a study carried out revealed that landforms such as lakes and fjords play a crucial role in the biogeochemical and global carbon cycle [18−20].

Major, trace, and rare earth element (REE) concentrations in combination with textural and mineralogical analysis of sediments have been used to understand geochemical fractionation in the sediment due to weathering and subsequent transport in the Svalbard region. Surface sediments from different water depths and a short sediment core have been collected from two fjord systems Krossfjorden and Kongfjorden. Bimodal distributions of grain size suggest that the finer size sediments were possibly deposited from the suspended load and coarser size from melting of sea ice and iceberg. The geochemical study reflects that the variation of the chemical composition of the sediments is due to first-order fractionations of elements during the grain size reduction and subsequent transport. The chemical weathering in the catchment area is incipient and the grain size reduction is dominated by mechanical weathering. REEs are mainly controlled by the presence of less weathered aluminosilicate minerals in these sediments. The source rock composition of the sediments is intermediate between granitic and granodioritic, with negligible contribution from the mafic rocks [21].

2.10.1 Diisopropylnaphthalene in the surface sediments of an Arctic fjord

Surface sediment samples from Kongsfjorden were analyzed for polycyclic aromatic hydrocarbons (PAHs) to understand their source and process responsible for their deposition in the sediment. PAHs were detected between 72.91 and 318.73 ng/g in 2012 and 72.54–311.97 ng/g in 2013 showing a decreasing trend from the mouth of the fjord toward the glacier head. During these years, 2, 6-diisopropylnapthalene (DIPN) showed concentration ranging from 1.21 to 5.87 ng/g and 1.31 to 4.79 ng/g with an increasing trend from the mouth of the fjord toward the glacial outlet. This systematic increase of DIPN toward the glacier is observed during both the years of sampling that supports the possibility of DIPN from human activities at Ny-Ålesund and its surroundings. DIPN in the environment has not been thoroughly investigated and its present-day concentrations may not be alarming; however, considering the increasing anthropogenic activities at Ny-Ålesund, it might be prudent to exercise caution to ensure that the levels do not increase over time [22].

2.10.2 Study on nonpolar isolates in the Tundra and fjord environment

The Arctic environment is warming twice as fast as the global average, which is affecting the biotic communities in a significant manner. Over the last 10 years, Indian researchers have been analyzing the changes in the microbial communities through culture-dependent and culture-independent (metagenome-based) techniques and evaluated some of the functional roles played by them. Our results reveal an increasing presence of nonpolar isolates in the Arctic environment (Fig. 13.4) that could be considered as a proxy and supporting evidence to the enhanced warming trends undergoing in the arctic [23].

2.10.3 Assessment of humic acids isolated from diverse Arctic environments

The nature and characteristics of humic acids isolated from the sediments of fjord, lake, and river environments in the Arctic region were studied for the biochemical composition and indices providing information regarding the productivity and nutritional quality of these sediments (Figs.13.5–13.8). The humic acids were isolated from each station and further analyzed for studying their elemental composition, structural characteristics, and the metal complexation properties with the help of an elemental analyzer and various spectroscopic techniques such as UV-Visible, FT-IR, and ICP-OES. The degree of carboxylation and lignin contributions toward the formation of humic acids is supported with various elemental ratios. The degree of aromaticity and various other specifications of the humic acid moieties were understood with the ratios obtained from the UV-Visible spectroscopic data. FTIR spectrums

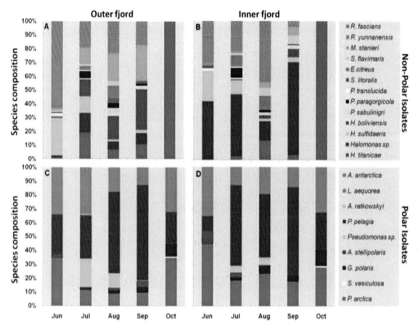

FIGURE 13.4 Distribution of polar and nonpolar isolates in the inner and outer fjords (Kongsforden) in 2011 during June–October. *Source: A.A. Mohamed Hatha and K.P. Krishnan, personal communication.*

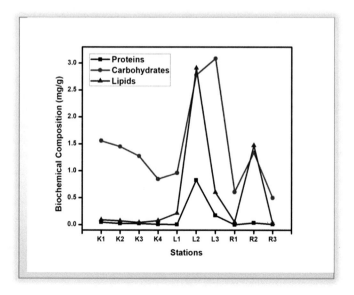

FIGURE 13.5 Biochemical composition of sedimentary organic matter obtained from the fjord, lakes, and river. *Source: Aswathy Shaji and Anu Gopinath, personal communication.*

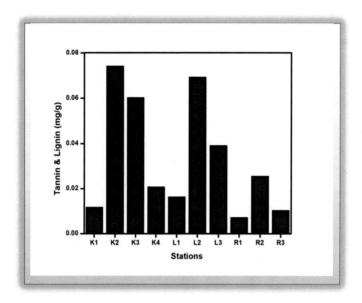

FIGURE 13.6 Tannin and Lignin component in the sediments of fjord, lakes, and river. *Source: Aswathy Shaji and Anu Gopinath, personal communication.*

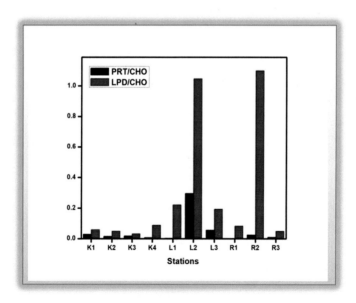

FIGURE 13.7 PRT/CHO and LPD/CHO ratios of the sediment organic matter from fjord, lakes, and river. *Source: Aswathy Shaji and Anu Gopinath, personal communication.*

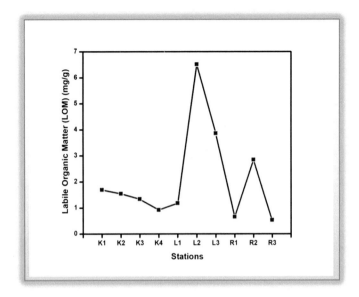

FIGURE 13.8 Labile organic matter (LOM) in the sedimentary organic matter of fjord, lakes, and river. *Source: Aswathy Shaji and Anu Gopinath, personal communication.*

provided an insight into the functional groups present in each humic acid isolated giving its uniqueness from the other. The metal complexing properties of humic acids were also a major part of this study with ICP-OES spectroscopic technique to distinguish the metals present in bulk and trace quantities within the humic acid [24].

2.11 Study on cyanobacteria

Cyanobacteria are interesting components of all polar habitats with many of them being present in different microhabitats. A long evolutionary history and strong adaptive tendencies allow these cyanobacteria to survive and thrive in these tough conditions. Polar cyanobacteria are interesting but challenging samples due to many reasons. The diverse array of microhabitats they dwell in and the successful adaptive strategies that they employ to thrive in these tough environments are remarkable. Indian researchers have initiated a detailed study on the cyanobacteria of the Arctic region (Fig. 13.9A and B) being in a pristine environment. But the challenge faced during sampling and post-sampling handling of cyanobacterial samples is yet another challenge. Loss of samples in the case of studies based on culturomics is a routine process and has severely masked the proper exploration of cyanobacterial cultures from these extreme habitats. The dearth of new cyanobacterial taxa being discovered from the Polar Regions can be simply understood by the fact that one of

FIGURE 13.9A Algal growth across different habitats in the polar regions: (A) typical algal growth in stagnant water bodies; (B) floccose algal growth in water bodies; (C) algal growth in small periodic streams below glacial runoff; (D) algal growth in long-running streams having stones and pebbles; (E) ball like algal growth on stone tops; (F) reddish colored algal growth on stone tops in running streams. *Source: Prashant Singh, personal communication.*

the most prominent and dominant cyanobacterial genus *Nostoc* has had a number of morphotypic genera described from various parts of the world [25−29], but not even a single new taxon has been described and discovered from the Polar habitats. It is envisaged that better strategies that could mimic the polar conditions holistically must be employed to gather better insights about the diversity and taxonomic complexity of polar cyanobacteria [30].

2.12 Deciphering the changing climate and environment around Ny-Ålesund, Svalbard

Reconstruction of the past climatic and environmental changes around Ny-Alesund, Svalbard has been one of the many mandates of the India Arctic Program. According to Indian researchers, [31] adopted a multiproxy approach involving investigations on palynology (Fig. 13.10A−J), quartz grain micro-texture, and magnetic susceptibility on the surface and subsurface sediments coupled with the examination of airborne pollen and polleniferous material

(b)

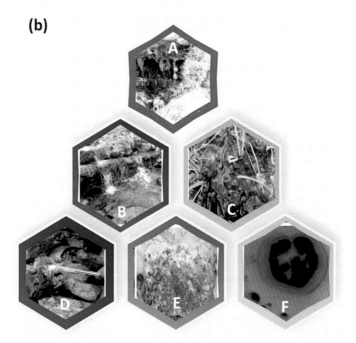

FIGURE 13.9B (A) shady hidden crevices in running glacial streams; (B) luxuriant algal growth in running streams; (C) characteristic growth of Nostoc commune complex in meadows; (D) algal growth beneath the ice in frozen streams; (E) Cyanobacterial colonies (most probably *Gloeo-trichia/Rivularia*) on stone tops in water bodies; (F) A typical Cryoconite. *Source: Prashant Singh, personal communication.*

from the flowers of the region. Their study concluded that Quartz grain microtextures were useful in reconstructing the paleoenvironmental conditions and reveal a predominance of glacial signatures, suggestive of the prevalence of overall fluvioglacial activities since the LGM. Although a broad similarity could be observed between the magnetic susceptibility and quartz grain microtexture inferences, nonetheless, these are not completely synchronous. According to them due to the paucity of palynomorphs in the subsurface sediments, vegetation-based paleoclimatic reconstructions could not be achieved yet but baseline data is now in place for future endeavors in this direction.

The inference drawn about past climatic reconstruction [28] includes the following:

- After the culmination of the LGM, during late Pleistocene (19,130–10,860 cal year B.P.), a warming phase is observed with evidences of glacial retreat.

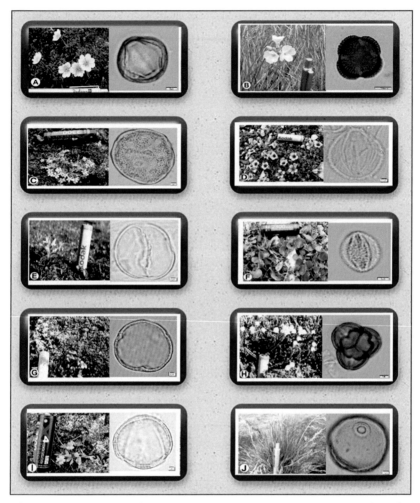

FIGURE 13.10 (A−J) Photographs of some flowers in the study area and their respective pollen. (A) *Dryas octopetala* (Rosaceae). (B) *Cardamine nymanii* (Brassicaceae). (C) *Arenaria pseudo-frigida* (Caryophyllaceae). (D) *Saxifraga oppositifolia* (Saxifragaceae). (E) *Pedicularis hirsute* (Orobanchaceae). (F) *Salix polaris* (Salicaceae). (G) *Oxyria digyna* (Polygonaceae). (H) *Cassiope tetragona* (Ericaceae). (I) *Ranunculus pygmaeus* (Ranunculaceae). (J) *Deschampsia alpina* (Poaceae). *Source: Ratan Kar, personal communication.*

- As the glaciers receded, the deglaciated area was taken over by the sea. This is well recorded by the presence of shell pieces in the lower part of the trench (70−100 cm). Aqueous markers in the quartz microtextures also provide evidences of the marine incursion.
- The Early Holocene period (10,860−8100 cal year B.P.) is marked by continuing warming. The further recession of glaciers resulted in an overall

influx of terrigenous sediments by glacial melt-water streams, which were deposited over the marine sediments.

- The recession of the glaciers is also marked by the presence of aeolian markers. As the area got devoid of the ice sheet, the aeolian activities too got registered in the depositional regime. The presence of a thin organic-rich band in the trench, dated 8100 cal year B.P., further confirms the subaerial exposure during this period.
- During Middle to Late Holocene (8100 cal year B.P. to the Present), a general fluvioglacial environment in and around the study area is observed, with glacial melt-water streams crisscrossing the strand flat and flowing into the fjord.

2.13 Exploring teleconnection between Arctic climate and tropical indian monsoon

The climate change over the Arctic region and North Atlantic shows a mechanistic link with the Indian Summer Monsoon (ISM) during the Holocene. The marine and continental archives of ISM precipitation suggest significant shifts during the Holocene aligned with the Arctic climate over multi-time scales. The ISM strengthened during the Greenlandian (11.7–8.3 kyr BP), showing variable but overall, decreasing precipitation during the Northgrippian (8.3–4.2 kyr BP), although synchronicity exists in paleoclimatic records owing to possible age errors and resolution, and proxy response to the changing climate. During the Meghalayan age (4.2 kyr to Recent), the Indian subcontinent witnessed a protracted dry event beginning at ~4.2 kyr BP and ended at ~3.4 kyr BP. Other significant events of the Meghalayan age include the Medieval Climate Anomaly and the Current Warm Period showing a strong ISM, interrupted by the Little Ice Age—a cold phase with low precipitation in the Indian subcontinent. The millennial-scale variability in the ISM is associated with the Heinrich and Bond events. The cooling in the Arctic Sea, ice expansion in the North Atlantic, and weakening of the Atlantic overturning meridional oscillations due to high freshwater flux and ice rafting in the North Atlantic caused weak ISM precipitation over the south and southeast Asia [32].

2.14 Assessment of black carbon aerosols and solar radiation over Himadri, NY-Ålesund

India has ventured into assessing the Black Carbon (BC) and measuring solar spectral at "Indian Arctic Station, Himadri," Ny-Ålesund, during 2011–14 (Fig. 13.11). The contribution from long-range transport of pollutants from faraway places is found to dominate the local sources such as emissions from shipping and power plant to the annual cycle with maximum BC mass

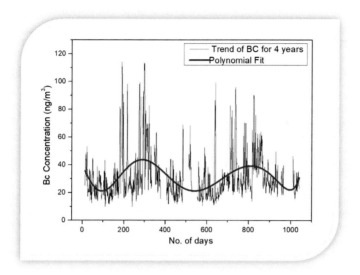

FIGURE 13.11 Long-term variations in BC mass concentration during 2011−14. Red curve indicates the polynomial fit to the data. *Source: S.M. Sonbawne, personal communication.*

concentration during winter/early-spring season and minimum during summer season. Moreover, higher BC concentrations were observed during 2012 as compared to other years during the study period. The spectral variations of aerosol optical depth observed during the summer months indicate the larger contribution of fine-mode aerosol particles to the BC mass concentration, particularly during 2012. Further, the zenith skylight spectra in the spectral range of 200−1100 nm indicate maximum particle scattered intensity around 500 nm. These results are found to play a vital role in the earth-atmosphere radiation balance and hence exhibit profound influence on regional and global climate change [33].

2.15 Assessment of Arctic geospace and space weather

A detailed investigation of Arctic Geospace and its space weather has been undertaken by Indian researchers. Some of the salient findings (Fig. 13.12A−D) are the following:

- The value of Total Electron Content (TEC) remains very high during the daytime as compared to that of night-time values at the low latitude region.
- The TEC variability shows the diurnal variation with the peak during afternoon hours and the minimum around premidnight hours at the low latitude region.
- Typical low latitude diurnal variation is absent at high latitude region and the variability of TEC through-out the day remains very less.

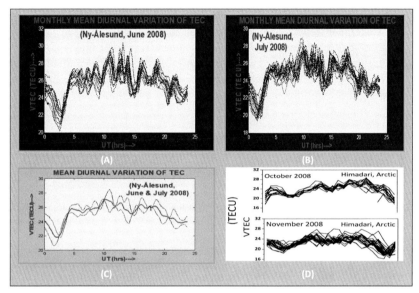

FIGURE 13.12   (A) Monthly mean diurnal variation of TEC in June 2008, (B) Monthly mean diurnal variation of TEC in July 2008, (C) Mean diurnal variation of TEC in June–July 2008, (D) Mean diurnal variation of TEC in October–November 2008. *(A–C) Source: A.K. Gwal, personal communication. (D) Source: A.K. Gwal, personal communication.*

- The time of occurrence of minimum value of TEC keeps on shifting as has been found (during the study of campaign mode data) in the month of February.
- The variability of TEC is high during the daytime and is comparatively low during night-time at the low latitude region.
- The behavior of quiet day TEC is different from that of the disturb days. The value of TEC remains high during the disturb days than that of the quiet day periods.
- A sharp increase in the value of Rate of TEC (ROT) has been found during the sunrise time and sunset time. Daily average TEC value is more correlated to the daytime average value of TEC than to the night-time average TEC values that mean some other physical processes also go during the night.
- The weak scintillations (S4 > 0.1) are observed all the 24 h of the day in almost all the seasons whereas during morning and afternoon hours slightly higher magnitude scintillations (S4 < 0.5) are also observed during the solar minimum period of 2008.
- Season-wise maximum occurrence is noted during summer months whereas in the winter and equinox months scintillations are observed mostly in early morning hours as well as in the night hours.

- As regarding month-wise maximum occurrence is observed between June and December months. At the starting of the month in this year, the scintillation activity is starting to grow from June when polar night starts.
- During the month of August and September, sun rises with minimum solar activity so that the ionization process starts again, but after the month of October, the scintillation activity again goes to minimum [34,35].

References

[1] S. Nayak, Balancing development and environmental concerns in the arctic, in: V. Sakhuja, K. Narula (Eds.), Asia and the Arctic: Narratives, Perspectives and Policies, Springer, Singapore, 2016, pp. 27−32.

[2] M. Menon, S. Dikishit, India Gets Observer Status in Arctic Council, The Hindu, 2013. https://www.thehindu.com/news/international/world/india-gets-observer-status-in-arctic-council/article4717770.ece. (Accessed 5 January 2020).

[3] S. Rajan, K.P. Krishnan, India's scientific endeavours in the arctic, in: V. Sakhuja, K. Narula (Eds.), Asia and the Arctic: Narratives, Perspectives and Policies, Springer, Singapore, 2016, pp. 43−48.

[4] http://www.ncaor.gov.in/arctics).

[5] S. Shetye, R. Mohan, S.K. Shukla, S. Maruthadu, R. Ravindra, Variability of Nonionellina labradorica Dawson in surface sediments from Kongsfjorden, west Spitsbergen, Act. Geol. Sinica-English Edition 85 (2011) 549−558.

[6] R. Saraswat, C. Roy, N. Khare, S.M. Saalim, S.R. Kurtarkar, Assessing the environmental significance of benthic foraminiferal morpho-groups from the northern high latitudinal regions, Pol. Sci. 18 (2018) 28−38.

[7] R.K. Sinha, K.P. Krishnan, A.A.M. Hatha, M. Rahiman, D.D. Thresyamma, S. Kerkar, Diversity of retrievable heterotrophic bacteria in Kongsfjorden, an Arctic fjord, Braz. J. Microbiol. 48 (2017a) 51−61.

[8] R.K. Sinha, K.P. Krishnan, S. Kerkar, D.D. Thresyamma, Influence of glacial melt and Atlantic water on bacterioplankton community of Kongsfjorden, an Arctic fjord, Ecol. Indic. 82 (2017b) 143−151.

[9] D.S. Singh, C.A. Dubey, D. Kumar, R. Ravindra, Climate events between 47.5 and 1 ka BP in glaciated terrain of the Ny-Alesund region, Arctic, using geomorphology and sedimentology of diversified morphological zones, Pol. Sci. 18 (2018) 123−134.

[10] R. Kar, A. Mazumder, K. Mishra, S.K. Patil, R. Ravindra, P.S. Ranhotra, P. Govil, R. Bajpai, K. Singh, Climatic history of Ny-Alesund region, Svalbard, over the last 19,000 yr: insights from quartz grain microtexture and magnetic susceptibility, Pol. Sci. 18 (2018) 189−196.

[11] Kumar A., Yadav J. and Mohan R. Global Warming Leading to Alarming Recession of the Arctic Sea-Ice Cover: Insights from Remote Sensing Observations and Model Reanalysis (Communicated) − Personal Communication.

[12] Kumar A., Yadav J., Srivastava R. and Mohan R. Arctic Sea Ice Variability and Trends in the Last Four Decades: Role of Ocean-Atmospheric Forcing (Communicated) − Personal Communication.

[13] Luis A.J., Mahanta K.K. and Shridhar D. Spatiotemporal Variability of Snowmelt Onset Across Svalbard Inferred from Scatterometer Data (2000-2017) (Communicated) − Personal Communication.

[14] Soheb M. and Ramanathan A. Mass Balance and Spatio-Temporal Change in the Area of Vestre Broggerbreen Glacier Ny-Ålesund, Svalbard, Arctic, between 1993-2018 (Communicated) − Personal Communication.

[15] http://www.ncaor.gov.in/pages/researchview/19.

[16] M.P. Raju, P.D. Safai, S.M. Sonbawne, C.V. Naidu, Black carbon radiative forcing over the Indian arctic station, Himadri during the arctic summer of 2012, Atmos. Res. 157 (2015) 29−36.

[17] https://punemirror.indiatimes.com/pune/others/arctic-pollution-in-2011-was-the-highestindecade/articleshow/45061604.cms.

[18] S. Choudhary, G.N. Nayak, N. Khare, Provenance, processes and productivity through spatial distribution of the surface sediments from Kongsfjord to Krossfjord system, Svalbard, J. Indian Assoc. Sedimentol. 35 (1) (2018) 47−56.

[19] S. Choudhary, Multiproxy Paleoclimate Reconstruction from the Sediments of High Latitude Regions, Goa University, India, 2019. Unpublished Ph.D. thesis.

[20] S. Choudhary, G.N. Nayak, N. Khare, Source, mobility, and bioavailability of metals in fjord sediments of Krossfjord-Kongsfjord system, Arctic, Svalbard, Environ. Sci. Pollut. Res. 27 (2020) 15130−15148.

[21] P. Kumar, J.K. Pattanaik, N. Khare, S. Balakrishnan, Geochemistry and provenance study of sediments from Krossfjorden and Kongsfjorden, Svalbard (arctic Ocean), Pol. Sci. 18 (2018) 72−82.

[22] N. Singh, S. Rajan, S. Choudhary, M. Peter, C. Krishnaiah, Diisopropylnaphthalene in the surface sediments of an Arctic fjord: environmental significance, Pol. Sci. 18 (2018) 142−146.

[23] Mohamed Hatha A. A. and Krishnan K.P. Increasing Presence of Non-polar Isolates in the Tundra and Fjord Environment − A Pointer towards Warming Trends in the Arctic (Communicated) − Personal Communication.

[24] Shaji A. and Gopinath A. Spectroscopic Characterization of Humic Acids Isolated from Diverse Arctic Environment (Communicated) - Personal Communication.

[25] K. Řeháková, J.R. Johansen, D.A. Casamatta, L. Xuesong, J. Vincent, Morphological and molecular characterization of selected desert soil cyanobacteria: three species new to science including Mojavia pulchra gen. et sp. nov, Phycologia 46 (2007) 481−502.

[26] P. Hrouzek, A. Lukešová, J. Mareš, S. Ventura, Description of the cyanobacterial genus Desmonostoc gen. nov. including D. muscorum comb. nov. as a distinct, phylogenetically coherent taxon related to the genus Nostoc, Fottea 13 (2013) 201−213.

[27] D.B. Genuário, M.G.M.V. Vaz, G.S. Hentschke, C.L. Sant'Anna, M.F. Fiore, Halotia gen. nov., a phylogenetically and physiologically coherent cyanobacterial genus isolated from marine coastal environments, Int. J. Syst. Evol. Microbiol. 65 (2015) 663−675.

[28] S.N. Bagchi, N. Dubey, P. Singh, Phylogenetically distant clade of Nostoc-like taxa with the description of Aliinostoc gen. nov. and Aliinostoc morphoplasticum sp. nov, Int. J. Syst. Evol. Microbiol. 67 (2017) 3329−3338.

[29] A.G. Saraf, H.G. Dawda, P. Singh, Desikacharya gen. nov., a phylogenetically distinct genus of cyanobacteria along with the description of two new species, Desikacharya nostocoides sp. nov. and Desikacharya soli sp. nov., and reclassification of Nostoc thermotolerans to Desikacharya thermotolerans comb. nov, Int. J. Syst. Evol. Microbiol. 69 (2019) 307−315.

[30] Singh P., Kumar N. and Pal S. Cyanobacteria in the Polar Regions: Diversity, Adaptation and Taxonomic Problems (Communicated) − Personal Communication.

[31] Kar R., Mishra A.K., Mazumder A., Mishra K., Patil S.K., Ranhotra P. S., Bajpai R. and Singh K. Deciphering the Changing Climate and Environment Around Ny-Ålesund,

Svalbard, Since the Last Glacial Maximum: A Multi-Proxy Approach (Communicated) - Personal Communication.

[32] Gupta A. K., Dutt S., Das M. and Singh R. K. Teleconnection between Arctic Climate and Tropical Indian Mansoon during the Holocene (Communicated) - Personal Communication.

[33] Sonbawne S.M., Devara P.C.S., Meena G., Saha S. K., Pandithurai G. and Safai P.D. Multi-year Measurements of Black Carbon Aderosols and Solar Radiation over Himadri, NY-Alesund: Effects on Arctic Climate (Communicated) - Personal Communication.

[34] Gwal A.K., Choudhary S. and Chaurasia H. Space Weather Phenomenon of Polar Ionosphere over the Arctic Region (Communicated) — Personal Communication.

[35] Gwal A.K., Choudhary S. and Bhawre P. Investigation of GPS Derived Total Electron Content (TEC) and Scintillation Index for Indian Arctic and Antarctic Stations (Communicated) — Personal Communication.

Chapter 14

Climatic change over Arctic

The Arctic is severely impacted by global warming compared to any other part of the world [1], resulting in a direct threat to the safety and livelihood of local inhabitants. During the past 30 years, Arctic sea ice melting is larger than Norway, Sweden, and Denmark combined. Because the Arctic helps to regulate the world's temperature, increased ice melting over the Arctic region would make the world even warmer. A close look at the factors that make the Arctic a driver of global climate change and associated warming revealed that Global warming is causing Arctic ice to melt, due to which the oceans around it absorb more sunlight and heat up, making the world warmer as a result as ice reflects sunlight, while water absorbs it.

1. The global warning and various feedback mechanisms

Climate feedbacks: processes are those that can either amplify or diminish the effects of climate forcing. A feedback that increases an initial warming is called a "positive feedback." A feedback that reduces an initial warming is a "negative feedback." The primary causes [2] and the wide-ranging effects [3,4] of global warming and resulting climate change are shown in Fig. 14.1A,B. Some effects constitute feedback mechanisms that intensify climate change and move it toward climate tipping points [5].

2. Temperatures changes over Arctic region

According to some estimates, the period of 1995−2005 was the warmest decade in the Arctic since at least the 17th century, with temperatures 2°C above the 1951−90 average [6]. Some regions within the Arctic have warmed even more rapidly, with Alaska and western Canada's temperature rising by 3−4°C [7]. The worrisome happened in the Arctic Circle has been that temperatures reached 38°C in a remote Siberian town, warmer than the previous temperature recorded in this region. GCMs have been used to project changes to the climate in a world with more atmospheric CO_2 since the 1990s. A common feature of these models is an effect called polar amplification. As a result, Arctic is getting more warmed [8]. This ice-albedo (really snow-albedo) feedback is potent in the Arctic because the Arctic Ocean is almost landlocked by Eurasia and North America, and it is less easy (compared to the Antarctic)

The Arctic. https://doi.org/10.1016/B978-0-12-823735-9.00005-9

219

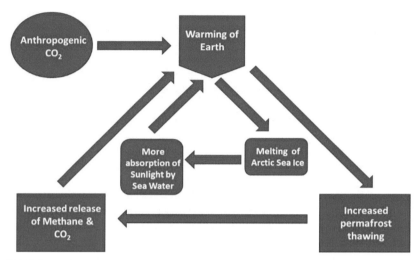

FIGURE 14.1A "Positive feedback". Mechanism over Arctic region responsible for Global warming.

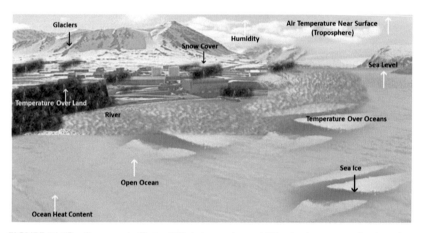

FIGURE 14.1B Causes and effects of Global warming and Climate change over Arctic region.

for ocean currents to move the sea ice around and out of the region. As a result, sea ice that stays in the Arctic for longer than a year has been declining at a rate of about 13% per decade since satellite records began in the late 1970s [8].

This warming has been caused not only by the rise in greenhouse gas concentration but also the deposition of soot on Arctic ice [9]. The current temperatures in the region have never been as high as recorded in proxy temperature records of the past. Perhaps, "anthropogenic increases in greenhouse gases have led to unprecedented regional warmth" [10,11].

3. Black carbon

Sailing ships generate Black Carbon deposits. Recent many-fold increase in such shipping operations in the Arctic region tend to increase Black Carbon that in turn reduce the albedo when deposited on snow and ice, and thus accelerate the effect of the melting of snow and sea ice [12]. According to an estimation about 60%, reductions in black carbon emissions and other minor greenhouse gases may lower down Arctic temperature (up to 0.2°C) by 2050 [13].

Further challenges, such as feedback loops and black carbon, are exacerbating this trend. Feedback Loops occur when the ice melts and the newly exposed surfaces both land and water absorb more of the sun's heat than snow or ice would. This in turn causes more ice and snow to melt, accelerating the process of climate change. Carbon emissions and other pollutants disproportionately affect the fragile Arctic climate. Black carbon pollution darkens the surfaces of snow and ice, causing them to absorb more heat and melt more quickly. It also can linger in the air, absorbing the sun's heat and causing temperatures to rise. Black carbon emitted north of 40°N in most cases in North America, Europe, or Russia has the greatest impact on the Arctic. The impact of the global warming over Arctic sea formation can easily be seen in Fig. 14.2A–D.

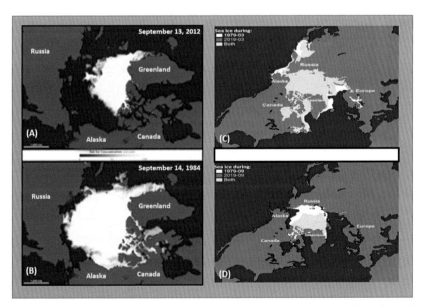

FIGURE 14.2 (A) Arctic sea ice areal extent during September 2012. (B) Arctic sea ice areal extent from September 1984. (C) Changes in the maximum sea ice extent for March spanning 1979–2019. (D) Changes in the minimum sea ice extent for September over 1979 to 2019 period. *(A) Source: https://commons.wikimedia.org/wiki/File:Arctic_Sea_Ice_Minimum_Comparison.png. (B) Source: https://commons.wikimedia.org/wiki/File:Arctic_Sea_Ice_Minimum_Comparison.png. (C) Source: A. J. Luis, personal communication. (D) Source: A. J. Luis, personal communication.*

Such a drastic change in areal sea ice extension in a short span (Fig. 14.2A,B) can easily be attributed to the recorded air temperature where average air temperatures (October 2010–September 2011) were up to 2°C more compared to the long-term average (1981–2010). Similarly, Fig. 14.2C,D shows changes in the maximum (March) and minimum (September) sea ice extent (SIE), since the advent of satellite-based measurements in 1979. These were processed by Ref. [14]. From an extent of 18.2 $\times$ 10^6 km^2 in March 1979, it dropped by 11.6%–14.52 $\times$ 10^6 km^2 in March 2019, which is the peak winter. Likewise, from September 1979, the SIE dropped by 42.6% from 7.05 $\times$ 10^6 to 4.32 $\times$ 10^6 km^2 for peak summer.

4. Arctic amplification

Polar regions of the Earth are more sensitive to subtle climate changes, and therefore the poles are warming faster than lower latitudes owing to the phenomenon known as "ice-albedo feedback" where more heating is generated by melting ice that tends to uncover darker land or ocean beneath (Fig. 14.3A,B), which in turn absorbs more sunlight [15,16]. The effect of sea ice loss on atmospheric temperatures and circulation patterns is of paramount importance because these changes will affect the terrestrial climate; therefore, the Land-surface snow is vulnerable to these changes [17]. The loss of the Arctic sea ice may represent a tipping point in global warming, these "tipping points" are thresholds where a tiny change could lead a system into a

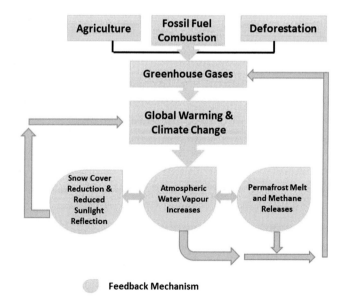

FIGURE 14.3A Schematic diagram showing ice-albedo feedback mechanism.

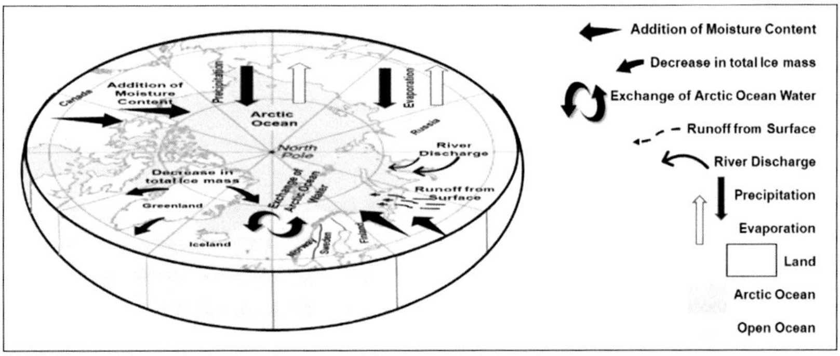

FIGURE 14.3B   Factors responsible for Arctic Amplification.

completely new state. Worldwide nine such tipping points are identified where a changing climate could push part of the Earth System into abrupt or irreversible change. It is expected that the impact of declining sea ice in amplifying surface air temperatures over the Arctic Ocean, that is, the "Arctic amplification" will become more pronounced as more sea ice is lost in the coming decades [17].

The pronounced warming signal, the amplified response of the Arctic to global warming, is often seen as a leading indicator of global warming. The melting of Greenland's ice sheet is linked to polar amplification [18,19]. According to an estimation, globally the sea level has gone up by approximately 4−8 inches and the sea level is expected to rise further ($\sim$23 feet) by 2100, due to continued Arctic sea ice melting. It may result in flooding of major coastal cities and submerge some small island countries. Furthermore, the Arctic being a storehouse of oil and gases, and this potential is being heavily exploited. Need not to say burning more fossil fuel will add up to the warming crisis. Perhaps, if we are successful in preventing drilling, in the Arctic region we may be able to protect the Arctic for its inhabitants.

5. Impact of climate change on Arctic

The Arctic region is facing unprecedented consequences in its weather due to globally occurring climatic changes; however, climatic changes are not solely responsible for the environmental changes in the Arctic. Other factors that contribute to the same are the dust of chemical pollutants from other regions, an extraordinary increase in the number of fishes harvested and other economic and social changes arising because of population explosion. These factors have been contributing to the environmental changes in the past and are anticipated to do so in the future as well.

The major credit to these changes in the Arctic region still goes to global warming; however, a very important fact in this respect is that the Arctic is itself contributing to these environmental changes at the global level. Let us try to understand the consequences of climatic change on the Arctic while assessing the resultant changes and effects of the Arctic climate change on other parts of the world.

6. Reasons for climatic change

It is an unbeatable fact put forth by geologists since ancient times that the earth's atmosphere is changing from time to time. Changes in the earth's orbit or rotation, increase and decrease in the solar energy received by earth's surface, changes in the orbit and activities of other planetary bodies of the universe, and eruption of volcanoes are some natural examples of changes in the earth's climate in the recent past. There are variations in the temperature of the environment, due to a rise in the temperature, the glaciers and ice of Polar Regions are melting at a constant rate and are moving toward the lower regions. The above-mentioned naturally occurring phenomenon has greatly been accelerated by human activities recently because of the blind race for development. These phenomena that were quite insignificant till recent past has now been exaggerated to such an extent that it is affecting human survival as well.

In the name of technological and scientific development and growth in the 21st century, human activities are accelerating the emission of greenhouse gases day by day. These gases that were playing an important role in maintaining the temperature of the earth to 14°C are now posing a threat to the survival and existence of plants, animals, and human life. From a boon, it has converted into a bane.

Greenhouse gases namely carbon dioxide, methane, nitrous oxide, ozone, and water vapor are always present in the atmosphere in trace amounts. 4% water vapor is present and is unequally distributed to which man has no control. The most prevalent among the greenhouse gases is carbon dioxide, which is responsible for increasing the temperature of the environment. Before the industrial revolution, the concentration of carbon dioxide in the atmosphere was 2.8 parts per 1000 parts. However, due to the combustion of coal and firewood and uncontrollable deforestation, its range has increased to 3.8 parts per 1000 parts and is still on a constant rise. This is the reason that carbon dioxide, which is a gas that sustains life on the earth is now becoming a life-threatening gas. It is said that presently the concentration of greenhouse gases consists of approximately 50% carbon dioxide. The concentration of other gases is compared with respect to the concentration of carbon dioxide in the atmosphere. Though methane and nitrogen oxide have a greater greenhouse effect than carbon dioxide their concentration in the atmosphere is quite low; therefore, they have a lesser contribution in terms of climatic change as compared to carbon dioxide. Although the Montreal Protocol put a restriction on the use of chlorofluorocarbon (CFC) that are depleting the ozone layer, still considerable quantities of CFC are present that are depleting the ozone layer putting the survival of plants and animals at stake.

During the past two centuries, the temperature of the atmosphere has increased by 0.6°C. Out of the total energy produced in the world, almost 80% of it is produced by burning firewood. Because of this and undue deforestation, the uncontrolled increase in the carbon dioxide concentration in the atmosphere cannot be curbed even by oceans and forests. Carbon dioxide in combination with other greenhouse gases protects the longitudinal waves from the outer space from coming into the earth's surface and is diverted back into the outer space; however, due to an increase in greenhouse gases, these longitudinal waves from the outer space are entering the atmosphere resulting in increased temperature. Although the temperature rise has shown its effect worldwide, yet the Arctic region has been doubly affected. Due to an ozone hole created above the Arctic region, because of the CFCs that destroy ozone, Alaska and East Canada have experienced an increase in 3−4°C in the winter season.

The effect of climatic change on the Arctic as compared to other regions of the earth has been exaggerated because of the following reasons:

i. Because of the melting ice from the Arctic, which forms layers in the oceans is of darker color, hence absorbs a greater amount of heat and energy resulting in the increase in temperature because of which further melting of ice occurs and the same process is repeated. This is a self-renewing phenomenon.

ii. Due to the greenhouse effect having a pronounced effect in the Arctic, the thermal energy entering from the outer space into the earth goes directly into the atmosphere instead of the earth's core. The earth's core further heats up the ocean resulting in the formation of water vapor thus enhancing the greenhouse effect.

iii. Heat energy from the sun directly reaches the earth's surface, having a greater effect on the Arctic because of the ozone hole over the Arctic region rather than other regions of the earth.

iv. Due to increased heating of the Arctic climate, the ocean water is receding drastically. As a result, the Arctic Ocean absorbs much heat during the summer and compensates for it during the winter.

v. Therefore, heat energy is being provided to the Arctic by the atmosphere and oceans as well. Because of these globally occurring changes, the climate of the Arctic is greatly being altered.

7. Arctic sea ice decline

Area, extent, and volume of the Arctic sea ice showing a drastic continued decrease and it is postulated that if such declining trend continues with the same pace by the 21st century there will not be any ice in the Arctic region. However, a neutral or positive effect on the nesting cycle of many Arctic-breeding shorebirds is expected from the ongoing warming of the Arctic region [20].

According to the report published in the Journal Natural Hazards [21], Ice in the Arctic sea is melting at an alarming rate, with its largest decline in 41 years being recorded in July 2019. The dramatic decline in Arctic sea ice due to global warming could well lead to the Arctic losing all its ice cover by 2050 if the melting rate does not slow down. The decline rate is very alarming. Based on the satellite data obtained from 1979 to 2019, the rate of surface warming and the changes in global atmospheric circulation have been studied for estimating the impact it had on the Arctic region [21]. Between 1979 and 2018, Arctic sea ice has been shrinking at a rate of 4.7% per decade, while the decline rate was found to be 13% in July 2019. An in-depth analysis of the satellite data of the past 41 years shows that the second-lowest sea ice extent was recorded on September 17, 2029 (4.1 million sq. km) as compared to the first lowest record of 2012 (3.3 million sq. km).

The study found that the process of sea ice formation during winters is unable to compensate for the sea ice that melts during the summer. The decline of sea ice has led to a localized increase in evaporation, air humidity, cloud cover, and rainfall, the study found. Warmer conditions meant that the Arctic sea ice froze later in the winter and melted sooner in the summer, which in turn directly affected not just local weather but the climate over the Arctic Ocean and peripheral regions. Moreover, the rapid decline in Arctic sea ice reported from the last 2 decades is due to climate change, but the state of sea ice in the spring and summer of this year is very critical [21]. Over the past 30 years, the Arctic has warmed at roughly twice the rate as the entire globe, a phenomenon known as Arctic amplification. Most scientists agree that the rapid warming and melting of the ice cover is a signal of climate change and global warming. Arctic sea ice is a critical indicator because the region helps cools the planet. The combination of ocean-atmospheric warming plays a lead role in sea ice melting processes. If these processes continue at this rate, then the existing 25% of sea ice in the summer months would be melted completely in perhaps another 12−15 years [21].

In another study conducted by Indian researchers highlights the increasing decline of the Arctic sea ice due to climate change. It reports that the September sea ice extent declined to 4.71 million square kilometers in 2018, its lowest in the past 4 decades [22]. The study also showed that September of 2018 was the third warmest on record, with temperature differences of the air above the Arctic Ocean ($\sim 3.5°C$) slightly higher than that of the Arctic land ($\sim 2.8°C$). The faster loss of sea ice for the whole Arctic Ocean during September demonstrates that there are substantial variations in surface air temperature, and there is a correspondence between the fluctuations in surface air temperature in the Arctic and global regions. The study suggests that due to more open seas in winters, the growth of sea ice is delayed, leading to disproportionate loss of sea ice occurring during summer [22].

8. Melting of the Greenland ice

In fact, there is evidence to indicate that sea ice extent has not been this low for at least the last 1500 years. Extreme melt events over the Greenland Ice Sheet (Fig. 14.4), which used to occur once in every 150 years, have been seen in 2012 and now 2019. Ice core data shows that the enhanced surface melting on the ice sheet over the past decade is unprecedented over the past three and a half centuries and potentially over the past 7000 years.

As per some estimation [23], Greenland will become warm enough by 2100 to begin an almost complete melt over the next 1000 years or more [24,25]. The projected predictive models indicate a sea-level rise of about 5 cm from melting of the expected during the 21st century.

According to a new study by researchers at Ohio State University, the ice sheet is retreating in rapid bursts, leading to a sudden and unpredictable rise in sea levels, making it difficult to prepare for the effects [26].

Based on the 4 decades of satellite data to measure changes in Greenland's ice sheet it has been observed that after 2000, the ice sheet shrank so rapidly that, and fresh snowfall would not replenish it. About 200 odd glaciers that make up the Greenland ice sheet have been observed retreating within the

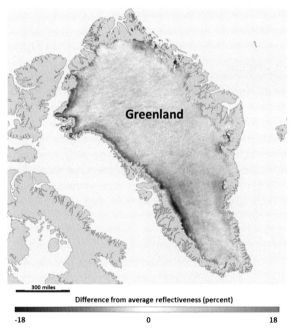

FIGURE 14.4 Drastic reduction in Ice-sheet of the Greenland due to continuous melting of ice owing to rising temperatures. *Source: https://en.wikipedia.org/wiki/Greenland_ice_sheet#/media/ File:Greenland_Albedo_Change.png.*

same episode. Ice melting in Greenland contributes more than a millimeter rise to sea level every year, and that is likely to get worse. Sea levels are projected to rise by more than 3 feet by the end of the century, wiping away beaches and coastal properties. Ocean warming is causing massive ice sheet loss in Greenland and Antarctica, confirmed by NASA, that the ice sheet is retreating in rapid bursts, leading to a sudden and unpredictable rise in sea levels, making it difficult to prepare for the effect [26].

9. Sudden changes

Generally, climate changes occur slowly in nature; however, according to weather experts, sudden and drastic effects in climatic changes may also be a result of which the critical threshold of the climate has been crossed. Water may change to vapor because of its latent heat of vaporization. These sudden changes may be much destructive as compared to the regularly occurring changes over time. After the end of the last ice age, during the end of the freezing period, the climatic condition of Greenland resulted in a 5°C decrease in temperature. Because of the decrease in the temperature due to the freezing of ice, sudden natural and varying changes have been observed. Due to these sudden changes, cold winds and cyclones are prevalent in the Arctic that pose a threat to the survival of Arctic flora and fauna. As per North Atlantic and Arctic Oscillations, the temperature of the Arctic region increased considerably between 1930 and 1940 and decreased between 1950 and 1960.

10. Adverse effects

A well-known fact regarding climate change is that the rate of change is much more effective than the actual change that has occurred. Any kind of data tabulation that is done is managed in such a way that it clearly signifies its effect on the human population. For example, only when the ice layer of the polar region melts at a slow rate and the seashore weathering occur at a slow rate, the construction of roads and buildings can take place. Hailstorms and rainfall rates have increased by 8% in the past 100 years and occur comparatively more often during winters than autumn and spring.

10.1 Sea ice

The layer of ice on the Arctic Ocean and nearby seas has shown a decrease of 8% in its expansion rate during the past 30 years. Weather scientists have shown that climate change indicator is used to check the change in the properties of the ice layer of the sea. Mining activities in these areas result in the evolution of poisonous gases from the core of the earth, which cannot be sensed by the miners; as a result, the concentration of these gases increases in the human body. The Canary bird can sense these poisonous gases at very low concentration and starts screaming in its own voice, which is used as an alarm for such gases. Presently, the Arctic Ocean ice and that on the nearby seas are sensitive toward the atmosphere and the sea, which are being warmed due to global warming. As a result of increasing temperature, the atmosphere, the polar ice is melting, and an area has become ice-free which is larger than the sum areas of Norway, Sweden, and Denmark. The layer of ice on the surface of the sea has been reduced by 10%−15% because of which the temperature of water, sea waves, and the water vapor arising from the surface of water bodies is being affected. The residents of the Arctic consider this layer of ice as their "elixir for life" and are dependent on the Arctic flora and fauna for their food and clothing.

During summers when the ice expands, Eskimos hunt for fishes as a food. Eskimos can walk easily on hard and thick ice layers or by sledges to hunt for fishes and other food sources. However, due to the undue expansion of the Arctic ice, the flora and fauna have been widely affected. In the year 1967, 1981, 2000, and 2002 when St. Lawrence in Canada showed absolutely zero ice accumulation resulted in the undue death of innumerable seals and their young ones.

During the past half-century, there is a considerable decrease in sea ice in the Arctic region. It is assumed that by 2100, it will decrease by 10%−50% more. With the help of five Climatic models it has been assumed that by the end of the century, the expansion of the ice will reduce to a large extent during summers. Accordingly, to some models of climate change by 2100, the ice covering the Arctic region will completely vanish during summers.

In fact, with time, the ice is getting far from the coastal areas of the Arctic Ocean. The albedo of the sea level, which is uncovered due to the melting of ice is less than the albedo of the ice. It absorbs more solar energy resulting in raise in temperature. The rate of melting the ice increases and the ice-free area increases. This "cycle" increases the pace of melting and consequences.

10.2 Polar Bear

Although the Polar Bear is a terrestrial animal, yet the polar ice is extremely crucial for it as it gives birth to its young ones on the thick layer of polar ice and is dependent on seals for its food that thrives on ice. The female polar bear comes out with its young ones in the spring season, at the time of reproduction it does not feed on anything, thus after giving birth, when it feels hungry it largely feeds on seals to satisfy its hunger of past 5−7 months. Due to the reduction of polar ice, the number of seals is decreasing resulting in the corresponding decrease in the number of Polar Bears due to disturbance in the food chain. Due to the receding polar ice, the fat and blubber of the Polar Bear are decreasing resulting in weaker young ones, which overall is resulting in the undue death of Polar Bear.

During the past 2 decades, the number of adult Polar Bears in Hudson Bay has decreased considerably because of the decrease in the expansion of ice the newborn bears could not survive. As per the studies conducted during the period 1991–98, the average weight of newly born polar bear was less by 15% and there was a decrease in the birth rate. Because of increasing rainfall and hailstorms, the dens of the Polar Bears were destroyed, and this is one of the reasons for several deaths of newborn and the mother bears. Various species of seals, for example, Harp Seal, Spotted Seal, and Ringed Seal hardly come on land and they spend most of their life in sea ice. They stay, breed, and feed them on ice only. Because of the reducing expansion of ice, the seals are also affected adversely.

10.3 Sea Birds

Due to the receding ice layers, the Ivory Gull and Little Oak population is being affected. Due to this phenomenon, in the past 20 years, a 90% decrease in the number of Ivory gulls has been observed.

10.4 Walrus

The most fertile and nutritive and suitable cultivation land is that area near the seashore. Due to the receding shoreline, the production cycle of the sea is getting affected adversely. Because of this, it is affecting the animals like Walrus that thrive in those areas. They must wander to distant places in search of food.

10.5 Fishes

The higher yield in the Arctic region is that of Arctic Fish. Arctic fishes are a delicacy worldwide. However, due to the increasing temperature of the ocean and sea, the fish population is reducing drastically due to difficulty in adaptability. Some important fishes namely cod and Gering have shown increased yield and growth due to the expansion of polar ice. Between the year 1920 and 1960, the number of cods in West Greenland was shown to have increased. However, due to the increase in the cod population, the population of shrimps (which cod feds on) is decreasing considerably.

The Arctic, home to hundreds of ice-dependent species, could experience an uptick in commercial fishing as its waters become more hospitable. Soon it will not be only traditional Arctic fish that will be found in the region: studies from the United Nations Environmental Programme World Conservation Monitoring Centre indicate that fish stocks are migrating northward as ocean waters warm. Fishing in northern waters is big business.

Arctic fisheries do have the potential to pose legal challenges, however: 2.8 of the 14.0 million square kilometers of the Arctic Ocean are international waters, and 92% of them are not yet governed by regional agreements to regulate commercial fishing. To protect the fragile ecosystem and establish sustainable fishing regulations, the five Arctic nations agreed in February 2014 to place a moratorium on commercial fishing in international waters until adequate scientific research is conducted and international standards are implemented. Other countries, such as the United States and Canada, have already implemented, or are now adopting, national policies on future fishing activities.

10.6 Engineering: problems

Due to the atmospheric changes with respect to temperature change, an ample number of engineering problems arise. Due to the melting ice of Siberia, the cities that are constructed on the ice are being destructed at an unprecedented rate. The permanent ice layer in Northern Russia is home to many buildings and population, which is being disturbed due to the melting ice. In the past few years, the temperature of permafrost has increased by 20°C. Due to the permafrost and other climatic changes, other very important complex internal activities occur that are given in the following:

i. Forests protect the permafrost from letting and the increase in temperature is reduced. However, due to forest fires, and pests' forests get destroyed. Therefore, the melting of permafrost is accelerated.
ii. Some plant species specifically black spruce helps in the stability of permafrost.
iii. Certain small rivers, streams, and lakes that arise from the Polar ice also have a pronounced change due to the melting of polar ice.
iv. The stable permafrost and the flora of the Polar Regions have a very complex relation. The cultivation of moss in these areas absorb the energy from the sun and prevents the permafrost from melting.

10.7 Hindrance in conveyance/transportation

It is very well said that the conveyance and transportation in the Arctic region are not conducive during winters. The Tundra region of the Arctic is covered with hard thick ice that allows only sledges and snowmobile to travel on it. During the summer season, this ice layer becomes weaker and is liable to be broken making the region nonconductive to travel. For the past 30 years, vehicles could move around these areas for 200 days but now it can be allowed only for 100 days. Because of the following reasons, the research expeditions carried out in such regions for the study of oil and natural gas have reduced by 50%. Till now, we have discussed the effect of climatic change on the Arctic Region; furthermore, we will discuss its after effects and future consequences.

10.8 Potential impact

The impact of the rise in atmospheric temperature is significantly visible in every physical factor of the Arctic. The carbon dioxide and greenhouse gases stay in the environment for a long time without breaking down. Because of this, there is a rapid increase in greenhouse gases in the atmosphere layer surrounding the earth. Even if we stop the emission of greenhouse gases today, their impact will stay for a long time. In this way, the Environment of the Arctic will not be able to overcome the impact of climatic changes for the next 100 years. With the help of B-2 Emission Scenario developed by the Inter-Governmental Panel on Climate Change, the Arctic Climate Impact Assessment has developed five models of climate to make assumptions about the next 100 years starting from 1990.

According to these assumptions, there is a possibility of an average increase of 3−5°C in the temperature of land regions of the Arctic. This increase can be up to 7°C in the atmosphere of the Arctic. In the winter season, the temperature of land areas may increase approximately by 7°C, and the atmosphere above the Arctic Ocean may show an increase of 7−10°C. The possibility of a rise in temperature is more in the land areas of the Arctic located in North Russia that are near those seas where the ice is melting rapidly.

10.9 Rain and snowfall

The rate of evaporation in water bodies is higher due to the rise of atmosphere temperature resulting from climatic change. Due to this, the amount of rain and snowfall has increased. It is estimated that by the end of the 21st century, the amount of rain and snowfall will increase by approximately 20% in the Arctic region. This increase will be mostly in rainfall. It has also been estimated that in the summer season, the northern region of North America and Chukotka region of Russia, but it will reduce in Scandinavian. But in contrast during winters the whole region of the Arctic (except Southern Greenland) will experience greater rainfall. The greatest increase in rainfall will be in the coastal areas during winter and autumn. It is estimated that this increase will be approximately 30%.

10.10 Terrestrial ice

The average volume of the terrestrial ice of the Arctic region is 31,00,000 cubic kilometers. If it melts completely, the water level of all the seas will rise up to 8 m. During the past 30 years, the expansion of terrestrial ice has reduced to 10%. In the future, it is expected to decrease more during the spring season. It means the Snow Season period will decrease. The rivers will provide more freshwater to the Arctic Ocean. At present, glaciers of the Arctic region have started melting before time. Since 1960, the volume of glaciers and ice caps is constantly decreasing, and the rate of decrease was the most during 1990. Though during this period, there was an increase in the volume of ice in the glaciers of Scandinavia and Iceland. The credit for this growth goes to increased rain in these regions. The largest sheet of the ice in the Arctic is on Greenland. It is also melting rapidly. The scientists have found that from 1979 to 2002 there was an increase in the rate of melting by 16%. In other words, due to the melting of the ice sheet in Greenland, Sweden has become ice free. In 2002, the rate of melting of ice was very high. During that year 2000-m-high ice was melted. The rate of melting is the highest in the glaciers of Alaska. It is estimated that out of the total melting of glaciers all over the world, the credit of half part goes to Alaska.

In the context of the whole Arctic region, there was a considerable loss in volume of ice during 1961−1998. During this period, the volume of glaciers of North America decreased by 450 cubic km and in Russian glaciers the decrease was 100 cubic km. With the increase in the melting of terrestrial ice and glaciers and increase in the water of rivers, the amount of fresh water in the Arctic Ocean will increase. As a result, the circulation cycle of the Atlantic Ocean will get disturbed and the climate of northwest Europe will be affected. It will become colder during winters. The ice layers of permafrost land will also melt rapidly. The land will become marshy and the animals of the Arctic will be faced with the problem of food and shelter.

10.11 Increase in water level of seas

According to climate models because of the melting of Arctic glaciers in the coming 100 years, there will be a worldwide increase in water level by 4−6 cm. According to some people, if we calculate the rate of melting of Arctic Glaciers during the past 2 decades, by 2100 the water level will rise more than that.

In this context, if we assume the scenario of the next 100 years by that time the melted ice of glaciers and sheets of the Arctic region would have risen the water level of the seas much higher than expected? The climatic models have also indicated that the temperature of the atmosphere of Greenland has increased three times than average. There is a possibility that in due course the ice sheet of Greenland will melt totally. As a result, the water level of the sea will rise to 7 m. The direct and indirect consequences of the rise in water level can be long lasting and disastrous. Therefore, this needs to be discussed in detail. Due to the rise in atmospheric temperature, the density of water in the sea decreases. In other words, due to heat, the volume of water increases as it expands. Therefore, density decreases but the level of the sea increases. This was the reason for the rise in sea level in the past 100 years although the amount of water added due to melting of glaciers. The water level of the sea is increasing due to the melting and increase in volume.

Scientists have found that in the past 100 years the level of the sea has increased by 10−20 cm. The rate of increase in water level during 1990 was 3 mm per year but a few decades earlier it was approximately 2 mm per year.

This rate would have been different for each sea, but it would have been the highest for the Arctic Ocean. It has various reasons:

   i. The average increase in temperature is more in the Arctic region than other regions
  ii. The glaciers of the Arctic region are melting at a higher rate
 iii. In the Arctic Ocean, the quantity of water provided by the rivers is more in comparison to other oceans.

Although this ocean has comparatively less amount of water (1% of total seawater worldwide) yet it gets more amount of water from rivers than other seas. As a result, the water level of the Arctic Ocean is increasing at a higher rate. Many big rivers fall in the Arctic Ocean, and they provide more water to it. The volume of water being provided by these rivers in cubic km is Siberian Rivers Severna-105, Pechora-108, Yenisei-108, Ob-402, Lena-528, Kolyma-103, North America—McKenzie-281 and Yukon-203. An increase has been observed in the quantity of water of these rivers due to the melting of terrestrial ice. This increase is more during winters. It has been estimated that in the next 100 years the increase will be 10%−25% and the maximum increase will be during winters. The increase in water level will adversely affect the sloppy areas and can lead to the vole oceanic eruptions. There is also the possibility of the sinking of some islands of the Pacific Ocean (Marshal, Kiribati, Tuvalu, Tonga, Micronesia, etc. and the Maldives of the Indian Ocean are expected to sink in water. The Lakshadweep Islands can also be in danger).

In Bangladesh, more than 1 crore 70 lakh population leave in coastal areas below 1 m high from the sea level. The rise in the sea level can also affect them. The cities of South East Asia like Kolkata, Mumbai, Dhaka, and Bangkok located on low lands and delta regions may also go underwater. Apart from the above, due to the rise in sea level, the salinity of water will increase in the Bay and vole oceanic areas. The erosion of coastal areas with soft soil will increase and many life cycles will be destroyed.

10.12 Displacement of vegetation belts

The Arctic region is divided into three parts based on vegetation grown in them—Polar (Cold), Desert, Tundra, and the North Part of the Boreal Forest. The polar desert is the North Part of the Arctic where only Algae and Moss is found. It is covered by ice for the whole year. On its south is Tundra where tiny trees and shrubs are grown. On the south of Tundra, trees with sharp leaves are found; this is called Boreal Forest. Due to the climate changes, the temperature of the atmosphere will increase that will help the growth of vegetation. Due to this displacement, the Boreal forest will cover the whole area of the desert. It is a fact that due to the displacement of the vegetation, the ecosystem will also migrate with them. There are many factors of displacement of ecological systems due to climatic change—the consciousness and responsiveness toward climatic change, the intensity of the ecological system, their periodicity, and suitability of soil, moisture, etc. During this displacement, the animal species adapt themselves in comparison to the vegetation species especially the migratory animals like Caribou, which adapt themselves as per the changing climate. The period of displacement varies as per the climatic change. Where the soil and other factors are suitable, the displacement can take place by the end of the century. Where the factors are not suitable, the displacement may take more time. Due to the rise in sea level and displacement of vegetation belts, the expansion of the Tundra will decrease considerably more than in the last 21,000 years. The breeding areas of the terrestrial animals and birds will shrink. The endangered species will vanish and other species in the vast regions of the Arctic will become endangered.

In the Arctic region, many plants and animals' species have adapted themselves as per the extreme environment but in this process, they have lost their potential to compete with other species. It is assumed that these species will not be able to survive the changing conditions and compete with the new species grown due to climate changes. It is assumed that due to environmental changes, these species will shift up to 1000 km toward the North and displace the existing plant species of those areas. But the species in the Northern part of the Arctic will not shift to the north because of the Arctic Ocean and they might be destroyed in due course. The displacement has started, and some birds, fishes, and butterfly species are being affected by this. Due to the rise in temperature, the Lichens and Moss species will decrease. It will be important to mention here that the Arctic is a vast reservoir of Lichens and Moss. Approximately 600 species of Moss and 2000 species of Lichens are found there, which is not found anywhere in the world.

10.13 Impact on natural forests

During the past few years, due to different reasons, forests were cleared in all parts of the world. As a result, their expansion has decreased. Earlier the expansion of forests was more in tropical regions but due to the rapid cutting of trees, it has reduced. Now, the natural forests exist from Siberia to North America (in the Northeastern part). The countries with the largest forests are Russia, Canada, the United States, and Alaska. These countries contain 31% of natural forest regions in the world. Boreal Forests occupy 17% of the total land area of the earth. In due course with increased temperature, these forests might shift to the North. Boreal forests have been of economic and environmental importance. In Finland, Sweden, and Canada, Boreal forests contribute about 10%—30% in income from export. Other than this, the maximum part of freshwater flowing to the Arctic region accumulates in these forests. These forests are also the place for the reproduction of many other species like Wolf, Wolverine, and Caribou. Now, the harmful effects of environmental changes are also being experienced there. The growth of many species of plants has decreased but got increased in other regions. The attacks by various insect species and forest fires are destroying the forests. Due to the melting of inner layers of permafrost, the land is converting into marshy land and is damaging the plants.

During the recent studies in Siberia, it has been observed that even after the last snow age, till 8000−9000 years ago the whole Russian Arctic region was covered with forests. The presence of microorganisms in the tree simply indicates that at that time the climatic condition of these forests was much better. During the ancient time, the forest region continued to shift due to natural reasons. Although due to climatic changes, Boreal forests are shifting toward the Tundra region, it is possible that some species of plants may not grow properly. Various factors like quality of soil, flood, and waterlogging can be responsible for this.

10.14 Terrestrial animals

The terrestrial Arctic animal species include small herbivores like squirrel, rabbit, lemming, etc. to big animals like Caribou, Musk Ox, and Carnivores like Polar Bear, Fox, and Scavenges. Although Polar Bear is the symbol of the Polar region, it is a terrestrial animal but it largely depends on aquatic animals for its food. Therefore, in the context of climate change, it will be better to discuss about Polar Bear with Aquatic animals. As compared to tropical climate regions, the number of ecosystems in the Arctic region is quite less and their interdependence is comparatively high. Therefore, when due to climatic region one ecosystem gets displaced, another ecosystem dependent on it also gets affected. For example, Moss and Lichens are being affected by the increase in temperature. Moss and Lichens are important elements in the food chain of Caribou. They provide food to Caribou during winters. Therefore, the adverse effect on Lichens and Moss are also affecting the population of Caribou.

Reindeers also get food from this region and their females breed in this region. Therefore, the adverse impact of the melting of permafrost is also affecting the Reindeers. The population of Peary Caribou's population has been reduced from 26,000 to 1000 between 1961 and 1997. Peary Caribou is a species of Caribou with low height and white color found in Islands of Canada and West Greenland. The population of migrant Porcupine species of Caribou of the United States, and Canada have been reduced from 178,000 to 123,000 during 1989−2001. The natives of those areas are also facing a lot of difficulties due to this. The decrease in population of these species is also affecting the hunting animals like Wolf, Wolverine, and Fox. A major part of the population of Eskimos still depends on Reindeers for their food, shelter, and transportation. Therefore, the Eskimos are also indirectly affected by the decrease in Lichens and Moss.

With the increase in the temperature in the atmosphere due to the melting of permafrost, a new cycle has started. The layer of ice generated from this cycle covers the plants that are the major food for the herbivore's animals. Therefore, the Lemming, Musk Ox, etc. do not get enough food.

10.15 Aquatic animals

Sea ice is constantly getting impacted over the Arctic region owing to rising temperature. It has been noticed that the average Arctic temperatures have doubly increased as compared to the global average rate. Such an unprecedented change in the Arctic sea ice severely affects several species of Arctic animals—including polar bears and Arctic foxes.

10.16 Polar Bear

There is a considerable decrease in the number of Polar Bears due to a decrease in the expansion of Arctic ice. In this context, some people are of the view that with complete melting of the sea ice and consistent global warming, by the end of the century the species of Polar Bear will become extinct. The only option for this is that the Polar Bear adopts itself according to the changing environment. It could adopt a tropical lifestyle. He might come across problems like breeding with Greasily Bear or Brown Bear, competition with other species of Bears, interaction with human, etc. The extinction of the Polar Bear will lead to the loss of various ecosystems of the Polar region.

10.17 Seal

The Ringed Seal will be affected the most by the decrease in the expansion of sea ice because it is more dependent on ice for survival. It makes a burrow in the sea ice and breeds during Spring for which she needs solid ice. It is beyond imagination to think that after the total melting of sea ice, the seal will start living on land. It never comes on land and cannot breed in land areas. Similarly, the spotted seal, harp seal, etc. will not be able to survive in absence of sea ice. Therefore, their population will decrease considerably.

10.18 Walrus

Due to the increase in temperature and shifting of ice from the coasts, the Walrus will not be able to find food and, consequently, their population will also get affected.

10.19 Sea Birds

The adverse impact of the decrease in the expansion of sea ice will also be felt by the Sea Birds like Gull and Little Oak. The Ivory Gull spends most of her life on Sea Ice. She makes next in rocks and breeds there. She is protected by the hunters in them. She hunts many fishes from the split ice. With the shifting of the ice from the coasts, the Gull will have to travel a lot for food which will have an adverse impact on its population.

10.20 Impact on human health

The changes in the climate also express their harmful effect on the human in the Arctic region. These effects on the Arctic are not the same for all the human. The seriousness of the impact depends on many factors like consciousness about personal health and environmental changes but generally in comparison to urban the impact is more on rural regions. The reasons being:

a. Rural areas lack the availability of health services
b. The lack of capability of adaptation in the people

Along with this the people in rural areas a still dependent on hunting the animals for their survival. The climatic changes adversely affect living organisms directly and its indirect effects can be observed on human health. The change in climate especially the increased temperature of the atmosphere has some advantages or disadvantages as follows:

With the increase in temperature of the atmosphere, there will be a decrease in the patients suffering from a disease like hypothermia and frostbite that are developed due to extreme cold. Nowadays, in Arctic regions, the death rate is much higher in summer as compared to winter. But according to some, there is no direct relation between an increase in temperature with diseases and death. This relation is also affected by different factors, for example, during winter many people die because of respiratory infection resulting in Influenza. Other impacts due to the climate changes are Heat Stress and incidents due to extraordinary melting of ice and change in weather. Other than this, the change in climate results in problems related to drinking water and sanitation.

Presently, in some parts of the Circumpolar system, an increase in temperature of the atmosphere results in complaints of such diseases that were not prevalent earlier. The respiratory problem is most prominent among them. Human cannot work hard due to this disease. The harmful effects of climatic change can be seen in the distribution of fishes and forest animals. The fishes and animals, which were the regular diet of the Arctic people, are getting extinct. To find them, the people are facing many problems. As a result, they are adopting the food habits of western people because of which diabetes, cancer, and heart disease patients are increasing. The animals that are migrating due to climate change are grabbed by communicable diseases.

Human is also being affected by these diseases. The Encephalitis caused by West Nile Virus is one example. West Nile Virus which creates Encephalitis originates from Tropical areas of Africa. It spreads in Mosquitoes then in birds, mammals, and human beings. It was first noticed on the Eastern coasts of North America. This virus spreads very quickly and within the span of 3 years by 2002 it spread in 43 States of the United States and six states of Canada. The birds became a victim of this West Nile Virus. Some of the birds were migratory, and they took the virus to other regions. The mosquitoes spread it not only in birds but in mammals and humans.

The unique thing in this aspect is that this virus has adapted itself according to the North American mosquitoes, and it has spread among 110 species of birds and many mammalians including human beings. Various species of mosquitoes affected by this virus were already present in the Arctic and therefore it got support from them in spreading it. Due to the warming up of the Arctic, the number of natural calamities like storms, land erosion, and floods is increasing, and they are likely to increase in the future. Human health will also be affected by this. On one hand, because of the melting of permafrost and erosion of coastal areas, the shortage of freshwater is increasing; on the other hand, the problems of absorption of wastewater are arising.

10.21 Depletion of ozone layer

After the Montreal Protocol and many International Agreements after that, the use of Chlorophyll carbon and other ozone-depleting chemical was reduced. But till that time, these chemicals were got deposited into the upper layers of the atmosphere in large quantities that will continuously reflect their harmful effects for many decades to follow. The Arctic has been affected gravely by these chemicals and will continue to be affected in the future as well. Due to this, the ultra violet rays from space, especially the B Rays (wavelength 280−320 nm) have the most harmful effects.

Generally, in the Arctic region, the ozone depletion and consequential increase in ultra violet rays are not equal; at some parts, these are in large quantity and at some parts less. Similarly, during a certain period of the year, the quantity of ultraviolet rays increases. It is during the spring season when the living organisms in the ecosystems become excessively sensitive. The main reason for this is the temperature changes in the stratosphere. The depletion of ozone in the stratosphere is caused due to an increase in the concentration of greenhouse gases. Due to the increase in the concentration of greenhouse gases, the temperature of the troposphere is increasing but that of stratosphere temperature keeps decreasing. Due to this, Polar Vortex occurrences are increasing. These polar vortices help the creation of clouds with ice particles. The ozone-depleting chemical reactions occur. These polar vortexes separate the stratosphere above the Arctic region. Therefore, ozone cannot come there. Therefore, the deficiency of degraded ozone is not getting fulfilled and the deficiency will continue for the coming decades. In other words, the ultraviolet rays will continue their destruction in the Arctic region.

10.22 Harmful effects

The increase in the concentration of ultraviolet rays has harmful effects on human beings, other animals, plants, single-cell planktons, and construction material. The increase in the concentration of ultraviolet rays and especially the B-grade rays will result in an increase in eye diseases and skin cancer. The immune system of the body will weaken, which will result in infectious diseases. These rays degrade the immunity system, thereby making some viruses active, for example, Herpes Complex Virus. It increases the possibility of viral diseases. Due to the increase in ultraviolet rays, the chance of mutation also increases in plants. There is direct and indirect impacts of these rays in the phytoplankton with many aquatic plants and animals.

Although since 1979 every year the ozone is depleting at the rate of 7% per year, the rate of ozone depletion increases during spring. The impact on the living organism is the most. Though, the ozone level has decreased by 30% −35%. During 1997, the daily average level of ozone was 40%−45% below the normal.

There are other factors that affect the depletion of the ozone layer apart from the concentration of UV rays falling on the surface of the earth. These are the following:

- Clouded sky
- The angle where the solar rays fall on the surface
- The height of the place from the sea level
- The albedo of the level
- Concentration of air soles

In the Arctic region, the maximum UV rays occur during the spring and summer seasons because at that time the sun rays comparatively fall at a large angle into the surface of the earth. This angle is least during autumn and winter. Therefore, at that time, a large quantity of UV rays is scattered into the earth's surface. But these rays reflect because the surface of the earth is covered with ice. The UV rays that generally fall on the human body are exactly the half that is already faced by the body. Presently, in the Arctic region, the people must face 30% more UV rays as compared to the first generation. Such a high quantity of UV radiation may prove very harmful to human beings.

These rays mainly affect the eyes adversely. While moving on the ice, the eyes of the human have to bear the UV rays that reflect back from the ice. The quantity of rays rapidly increases when the ozone concentration on the stratosphere reduces during the spring season. The UV rays that fall while looking at the horizon are the same as that of looking at the sky. The effect caused by UV rays on plants and animals can be amended by climatic change. For example, the plants and trees grown on the ice-covered Arctic have to face the UV rays apart from the reflected rays of the ice. Due to the climatic change when the ice is melted, the reflection of rays will reduce. In contrast, the plants and animals living under the ice were still unaffected by these rays due to the protective sheet of ice. With the melting of ice, they will be in direct contact with these rays. Similarly, the animals that were under the ice of lake and rivers will also be uncovered and meet UV rays.

Climate change has a worldwide effect, but many studies show that warmer trends are more intense in higher latitudes. Since the Arctic resides in higher latitude, its communities are more vulnerable. Thus far, average temperatures have risen almost double as fast as in the rest of the world. Records show variations in numerous parts of the biophysical environment of the Arctic, such as sea-ice extent, area of permafrost and depth, river hydrology, geophysical processes, and the distribution of marine and terrestrial species of the Arctic. The warming of the earth's atmosphere and surface is mainly attributed to the greenhouse effect. The gases that have a role in the greenhouse effect are mainly carbon dioxide, methane, nitrous oxide, and CFCs. Even though carbon dioxide is an abundant greenhouse gas, other gases absorb radiation more efficiently and persist in the atmosphere longer than carbon dioxide, so their warming effects increase with time [27].

Changes in climate have made Arctic resources more accessible. For example minerals, such as rare-earth metals in Denmark have been exposed by retreating the ice caps. These metals can be extracted and used for technologies like cell phones or military guidance systems. In addition, climate change in the Arctic resulted in new trade waterways through the north, further exploiting the area. Changes in the Arctic will affect resource competition and conflict in the upcoming years and will have worldwide impacts. The Arctic basin is an ice-covered ocean that has strong feedback effects on many parts of the climate system. Since the arctic regulates heat exchange between the ocean and atmosphere, the sea ice decline is expected to affect atmospheric circulation and weather patterns. Other effects include diminished rainfall in many parts of the world, leading to desertification in many areas and a decline in their ability to sustain agriculture. These changes indirectly create more conflict in water scarcity and will increase migration of communities. More direct effects of the degradation of the Arctic include a significant rise in global sea levels, which will displace low coastal areas around the world and result in the loss of agricultural lands [28].

11. Effect of Arctic on global climate

If the changes in the world-wide climate affect the arctic region, the Arctic also affects the climate of other regions of the world. The major part of the Sun energy is consumed by the regions near the equator and the more we move toward poles the quantity of the energy reduces. Consequently, the quantity of sunlight and energy is the lowest on the Poles. A major part of the Polar Regions is covered by ice therefore a major part of the energy (80%–90%) is reflected by the ice. Certainly, the quantity of reflected solar heat in the polar region is greater than in the tropical region. If the environment and the sea do not absorb the heat in the tropics and do not transfer it to the poles, then the tropics will become more temperate (hot) and the poles will remain cold. The Atlantic Ocean performs the work of heat transfer in the Northern Hemisphere. On the other side, various activities taking place in the Arctic region itself affect the circulation of the Atlantic Ocean.

The activities occurring in the land and water of the Arctic region work in three ways:

1. Due to an increase in the quantity of greenhouse gases, reducing the expansion of ice on the Arctic and converting the ice-free areas into rich vegetation areas;
2. By melting the ice of Arctic in large quantity to add fresh water to the seas and thereby changing the circulation cycle of the Atlantic; and
3. To excrete the greenhouse gases in the large amount due to an increase in temperature of the environment.

11.1 Decrease in expansion of ice

Due to an increase in the amount of greenhouse gases in the environment, the increase in temperature results in the melting of ice in land and ocean on a large scale. Apart from this, the water takes more time in converting into ice during winters in the Arctic region, but during spring, melts rapidly. Due to this, the land and sea have become more ice-free in comparison to the past. The ice reflects 80%–90% of the solar energy due to its glittering white color but vegetation and seawater reflect solar energy in comparatively very less quantity, approximately 20% to 10%, and the remaining energy is absorbed. Due to this, the earth's surface becomes more hot or temperate. As a result, a chain reaction starts. More amount of ice melts on the hot earth. So, on the uncovered earth and sea, the expansion of vegetation and water increases, respectively. It starts absorbing solar energy in large amounts and due to which more ice melts (Fig. 14.2A–D). In this way, this chain reaction gets momentum itself. This sequential process has been started in the Arctic. As a result, glacier, land ice, and sea ice are melting in large quantity. Though these are regional events, the impacts are being observed in every part of the world. The rate of change of environment from hot to hotter is increasing. As a result, 10% of the Arctic land has become ice-free during the past 30 years. Consequently, due to the increased rate of becoming ice free, forests have developed even in those areas where there was no vegetation earlier.

In other regions of the world due to the increased burning of fossil fuels, the emission of carbon dioxide soot has been generated in large quantities. The wind carries off this soot into the Arctic region where it settles on the ice. As a result, the capacity of the ice to reflect solar energy decreases. It starts absorbing solar energy in large amounts and resulting in the melting of ice in more quantity.

11.2 Effect on ocean circulation

As a consequence of warming up of polar region, it has been stipulated that in the future there could be a shutdown of the thermohaline circulation (Fig. 9.8A,B) similar to that which is believed to have driven the Younger Dryas, an abrupt climate change event. There is also potentially a possibility of a more general disruption of ocean circulation, which may lead to an ocean anoxic event; these are believed to be much more common in the distant past.

11.3 Changes in the circulation of Atlantic Ocean

The events occurring in the Arctic affect the circulation of the Atlantic Ocean, changing the environment of northwest Europe directly. Because of global warming, the changes occurring in many natural causative factors are also affecting the circulatory system of the Atlantic Ocean. Water streams of the sea transfer some parts of the total solar energy received by the equatorial region to the Polar region. These water streams are produced due to the variability in heat and salinity. Gulf Stream produced from Bay of Mexico in the Atlantic Ocean is one of such water streams. It originates from the temperate region, flows toward Northeast, and reaches to Northwest European coast. It also carries heat in large amounts. Due to this heat, the temperature does not fall in Northwest European countries as much as in other regions situated at the same latitude even in winters. But the water of the Gulf Stream transfers its heat to other things and becomes cold. So, its density also increases. As a result, after reaching the Northern region of the Atlantic Ocean especially in the Labrador Sea, it settles down at the bottom. To fulfill the gap, the water from the temperate zone starts flowing toward the North. In this way, a "Thermohaline" circulation is established. This thermohaline circulation is also called as "Conveyer Belt." Details are provided in Chapter 9.

Another factor also helps in the above happening. When the water of the sea is converted into ice, the salts dissolved in it become separated in a large amount. It gets dissolved in the remaining water. Due to this, the concentration of salts increases in the remaining water and it moves toward the bottom of the sea becoming heavy. In this way, the heavy water going toward the sea leaves mixes with the cold heavy water of Gulf Stream in the Continental Shelf area of Arctic. Another activity also takes place with this. Some parts of the sea and land ice of the Arctic are also continuously melted. The resultant freshwater keeps floating over the heavy water like the oil on the water surface. Usually, a fine equilibrium keeps existing between the saline heavy water and the freshwater of the Atlantic Ocean due to which in the Atlantic Ocean the thermohaline circulation keeps moving appropriately. Because of changes in the environment, the ice and glaciers of the Arctic Ocean are melting more rapidly. Due to which fresh, comparatively light water is mixing in the Atlantic Ocean in large amounts. As a result, the longitudinal mixing of oceanic water and the development of deepwater are becoming slow. Consequently, the thermohaline circulation of the Atlantic is slowing down. Due to the slowing down of the thermohaline circulation, the amount of CO_2 to be absorbed by the ocean decreases and the concentration of CO_2 is increased in the environment resulting in a rise in temperature of the environment. Due to the lowering of the thermohaline circulation, the rate of streams of the Atlantic Ocean to carry the heat toward the North from the Equatorial region will also slow down. As a result, the temperature of Northwest Europe will fall. It will be a surprising situation in which the temperature of Northwest Europe may fall regionally whereas the temperature of other regions of the world is increasing.

Due to a decrease in the rate of formation of water at the bottom of the Arctic Ocean, the other oceans of the world will also be affected. The heat carried away on the surface by the factors moving upwards of thermohaline circulation and the amount of nutritive materials will also decrease. Due to which the rate of rise of sea-level will increase but the deficiency in nutritive material in the water of sea-surface will adversely affect the sea animals. The amount of carbon in the form of dead bodies of animal's moves toward the bottom from the surface will also be decreased due to the deficiency of these nutritive materials.

11.4 Excretion of greenhouse gases

The Arctic region is a potential storage of the greenhouse gases that get released due to ongoing Global warming accelerates its release.

11.5 Arctic permafrost thaw

Rapidly thawing Arctic permafrost (Fig. 14.5) due to higher temperatures release large amounts of carbon into the atmosphere. In the permafrost, there is twice as much carbon as in the atmosphere. According to an estimation, about 100 billion tons of carbon is expected to be released in the 21st century [29]. According to "Arctic report card" [30] (2019), the current greenhouse gas emissions from Arctic permafrost are almost equal to the emissions of Russia or Japan or less than 10% of the global emissions from fossil fuels.

In most parts of the Arctic, the layers of ice are accumulated, and the earth has become a permanently frosted land (Permafrost). A large amount of carbon is embedded in these layers in the form of organic materials. Carbon is

FIGURE 14.5 Permafrost thaw and coastal erosion on the Beaufort Sea, Arctic Ocean. *Source: https://commons.wikimedia.org/wiki/File:Beaufort_Permafrost1.JPG.*

accumulating in large amounts especially in the large water-logged marsh of Siberia and North America. During the summer season when the ice layers accumulated under the earth are being melted, the organic matters embedded in it are breaking down chemically. As a result, carbon dioxide and methane (Co_2 and CH_4) are being released. These increase the temperature of the environment. Due to the rise in temperature of the environment, the ice layers will melt in large amounts and consequently Co_2 and CH_4 will also be generated and released in large amounts and will dissolve in the environment. As a result, the temperature of the environment will rise significantly and melt the permafrost in large amounts. All this will create a cycle that might take rapid speed with time. The other factors like the moisture on the soil will also become active partners in this cycle. The scientists are not able to impacts and result of these activities.

In Boreal and Tundra forest areas of Arctic large amount of land, Carbon is available in the form of plants and soil. Dead plant materials are broken down in the soil of moist land and ponds of Tundras. Methane is released from it. Due to the rise in temperature and increased rainfall, the intensity of the environment, the rate of release of methane increases though the methane is absorbed by the land where the soil becomes dry. Carbon is also released in large amount due to the decomposition of plants and forest fires in dry parts of Arctic regions. Although in the beginning due to rise in temperature the rate of decomposition of the plant will increase but after some time due to excessive growth of vegetation and increased area of forests, the amount of carbon dioxide to be absorbed will also increase.

In fact, scientists are not able to confirm which activity will be more intense. According to some people due to a rise in the temperature of the environment, the increase in the expansion of forests will result in a decrease in the amount of carbon dioxide in the environment. On the permafrost and sediments of shallow parts of Arctic Ocean CH_4 (Methane), solid ice methane in the hydrated form is present in large amounts. If the temperature of permafrost of the temperature of the bottom of the shallow sea rises to some degrees, the breaking of methane hydrate starts. Methane will get released. Though the methane liberated by this method has comparatively less effect on environmental change but due to the rise in temperature of permafrost or seawater, the amount of methane liberated from methane hydrates may also increase to a large extent. It can affect its neighboring environment. The Arctic Ocean has no important contribution to the absorption of carbon dioxide from the environment yet. Its reason is that the maximum part of the Arctic Ocean is covered with ice. As a result of environmental change, a situation is coming where a big portion of the Arctic Ocean is getting snow-free. The snow deficient cold water of the Arctic will absorb carbon dioxide in large quantities at that time. Due to this, the bioproductivity of the Arctic Ocean may increase.

11.6 Arctic ocean acidification

Increasing seawater carbon dioxide concentrations leads to a relative reduction of pH of the seawater. This phenomenon is known as "Ocean Acidification." In other words, Ocean acidification refers to the reduction in the pH of the ocean over an extended period, typically decades or longer, which is caused primarily by uptake of carbon dioxide (CO_2) from the atmosphere but can also be caused by other chemical additions or subtractions from the ocean [31,32].

This factor is impacting marine organisms, ecosystems, and biogeochemical cycling. Arctic Ocean is experiencing the fastest rate of ocean acidification. Unless global carbon emissions are drastically curtailed, the ongoing marine carbonate system changes linked with significant shifts in ecological status over the large parts of the Arctic Ocean in the coming decades, are growing challenges including significant socioecological and economic consequences at the local to global level.

According to the AMAP-Arctic Ocean acidification report [32], the Arctic Ocean acidification has been affecting the Arctic marine environment and ecosystems in following manner;

- Arctic marine waters are experiencing widespread and rapid ocean acidification with the primary driver being uptake of carbon dioxide emitted to the atmosphere by human activities.
- The Arctic Ocean is especially vulnerable to ocean acidification, which is not uniform across the Arctic Ocean.

- Arctic marine ecosystems are highly likely to undergo significant change due to ocean acidification, with direct and indirect effects on Arctic marine life.
- It is likely that some marine organisms will respond positively to new conditions associated with ocean acidification, while others will be disadvantaged, possibly to the point of local extinction.

Intensified ocean acidification in the Arctic ocean is attributed to low temperatures, increased freshwater supply (river runoff and ice melt), and low pH Pacific water inflow. Seawater pH, the partial pressure of CO_2 (pCO_2) and the saturation states of aragonite (Ωarg) and calcite (Ωcal) [33]. They are major drivers of key marine physiological processes and are used to indicate potential challenges to some marine species [33].

11.7 Global effects of Arctic changes

The Arctic region plays an important role in maintaining a stable global climate, atmospheric systems, and oceanic circulation. Changes in the Arctic climate will affect weather patterns, temperatures, and biological diversity worldwide. For instance, Refs. [32,33] has found that melting glaciers and sea ice are the greatest contributors to the global rise in sea levels. In addition, thawing Arctic permafrost releases carbon dioxide and methane gases into the atmosphere, accelerating global climate change. Finally, the loss of wildlife resulting from these changes disrupts the global food chain, including the fish, birds, and mammals that provide important food sources for humans.

Since the 15th century, when the quest for a Northwest Passage to shorten shipping time to the Orient propelled men to risk their lives trying to cross the ice along modern-day Canada's borders the dream of an open shipping passage through the Arctic has enticed business and political leaders. As climate change continues to melt the Arctic ice, this dream and even the prospect of ice-free Arctic summers could become reality as early as the 2030s. However, the advancement of shipping and trade routes in the region will depend on the development of the security and geopolitical environment over the next 2 decades. In this increasingly complex operational landscape, the actions of governments, firms, and other organizations will shape the direction of future Arctic relations and business opportunities.

12. Social impacts of Arctic warming

Constant warning of the Arctic region is having a direct impact on the people that live in the Arctic such as Inuit, as well as other societies around the world [34].

Increasing temperature will adversely impact hunting, which is a major way of survival for some small communities [35], sea ice formation that will

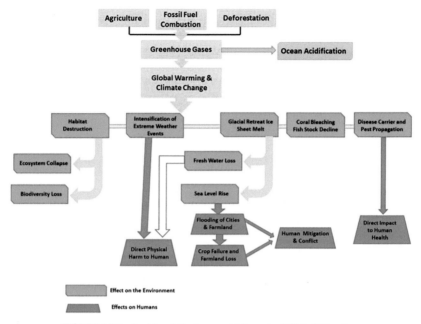

FIGURE 14.6 Social and Environmental impact of Global Warming.

cause certain species populations to decline or even become extinct, creating issues for Inuit hunters [35]. Similarly, unforeseen changes in river and snow conditions will cause herds of animals, including reindeer to change migration patterns, calving grounds, and forage availability [34].

Some transportation routes and pipelines on land being disrupted by the melting of ice [34]. Quality and characteristics of snow that is used to build shelters such as igloos have also not been untouched by the ongoing climate change in the Arctic. The changing landscape and unpredictability of the weather are creating new challenges in the Arctic [36]. The rapid Arctic climate change is forcing the indigenous people to adopt to entirely new living conditions in the Arctic requiring them to change their identity and diet as their tradition, culture, and knowledge are all developed within their ecosystems, Fig. 14.6 reflects the social and environmental impact of ongoing climate change affecting polar regions at faster rate affecting living beings and their habitats.

12.1 Territorial claims in the Arctic

Shrinking Arctic ice (Fig. 14.7A) has added to the urgency of several nations' Arctic territorial claims (Fig. 14.7B) in hopes of establishing resource development and new shipping lanes (Fig. 14.7C), in addition to protecting sovereign rights [37].

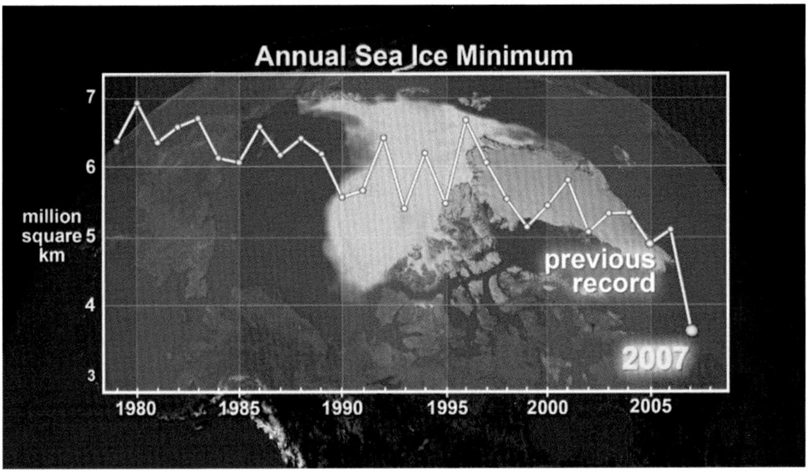

FIGURE 14.7A Annual Arctic sea ice minimum. *Source: https://upload.wikimedia.org/ wiakipedia/commons/9/97/Annual_Arctic_Sea_Ice_Minimum.jpg.*

12.2 Social and commercial movements to Arctic

Trans-Arctic Route, in contrast to the Northeast Passage largely avoids the territorial waters of Arctic states and lies in international high seas [38]. There is a burgeoning interest of Governments and private industry in the Arctic [39] shortened distance of these northern routes and access to natural resources including valuable minerals and offshore oil and gas [34] will be an added advantage to many commercial establishments, industries, and companies. Nevertheless, managing and controlling these resources will be quite challenging owing to continually moving ice [34]. However, with the opening of the more ice-free area in the Arctic will be a boost to Tourism industries as less sea ice will improve safety and accessibility to the Arctic [34].

13. Paradox

On the one side due to the climatic change in the Arctic region the forests are shifting toward the north; on the other hand, the possibility of expansion of desert cannot be ruled out. This is a paradox. This condition is created due to the assumptions of some scientists. Some scientists have assumed that increase in temperature in the climate will help in the expansion of forests in the Arctic and reduce carbon dioxide from the atmosphere, but they also mentioned that many factors will contribute to this process that is not yet known. Sometimes the contribution can be negative also. The end results can be the opposite of the predicted ones. In other words, a rise in temperature can lead to the expansion of desert in some areas of the Arctic. A doubt is coming up in respect to the conclusions of recent studies that the rise in temperature due to

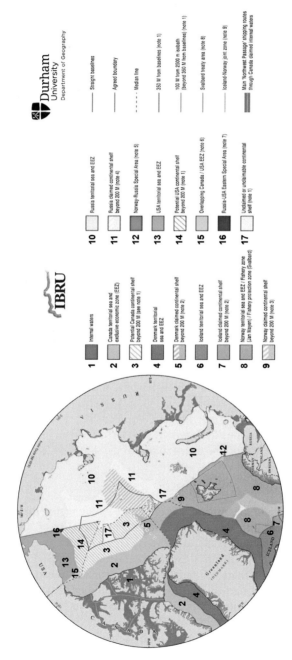

FIGURE 14.7B Arctic territorial claims. *Source: https://www.sciencealert.com/this-map-shows-all-country-s-claims-on-the-arctic-seafloor.*

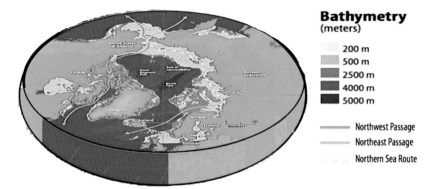

FIGURE 14.7C   Various Arctic shipping routes. *Source: https://en.wikipedia.org/wiki/Arctic_ shipping_routes#/media/File:Map_of_the_Arctic_region_showing_the_Northeast_Passage,_the_ Northern_Sea_Route_and_Northwest_Passage,_and_bathymetry.png.*

more absorption of solar energy by the expanded forest region, the increase in solar energy might be more than the increase in carbon dioxide resulting in more rise in atmospheric temperature.

Although it is being concluded that the rainfall will increase due to the increase in temperature in Arctic region, we cannot be sure that it will happen in the whole Arctic region in all seasons because of the rise in temperature the evaporation rate increases. When the rain cannot compete with the increase in evaporation, then the earth becomes dry and converts into a desert in due course.

In this context due to the melting of permafrost, the problem of flow of the water is created. In summers the upper layer of permafrost melts which makes the earth moister and smoother. But when innermost ice layer starts melting, then this smoothness may disappear. As a result, in most parts of Arctic, especially those regions where there were no glaciers 10,000 years ago and which have thin particles of soil brought by the air on its permafrost, will become dry and will be eroded. The ancient records of climate change show that such changes occurred in cold and dry grasslands of Siberia and Alaska. Due to the above activities due to increase in temperature the forest is expanded but after some time some area gets dried and convert into desert.

References

[1]   J.J. McCarthy, Climate Change 2001: Impacts, Adaptation and Vulnerability. Contribution of Working Group II to the Third Assessment Report of the Intergovernmental Panel on Climate Change, Cambridge University Press, New York, 2001. ISBN 978-0-521-80768-5. Archived from the original on 14 May 2016. Retrieved 24 December 2007.

[2]   The Causes of Climate Change climate.nasa.gov. NASA. Archived from the original on December 21, 2019.

[3] Climate Science Special Report/Fourth National Climate assessment (NCA4), Volume I science 2017. globalchange.gov. U.S. Global Change Research Program. Archived from the original on December 14, 2019.

[4] Summary for Policymakers, Intergovernmental Panel on Climate Change, 2019, p. 6.

[5] The Study of Earth as an Integrated System. nasa.gov. NASA. 2016. Archived from the original on November 2, 2016.

[6] R. Przybyla, Recent air-temperature changes in the Arctic, Ann. Glaciol. 46 (2007) 316−324, https://doi.org/10.3189/172756407782871666.

[7] Arctic Climate Impact Assessment 2004. Arctic Climate Impact Assessment. Cambridge University Press, ISBN 0-521-61778-2, siehe online Archived 28 June 2013 at the Wayback Machine.

[8] J. Bamber, Siberia Heatwave: Why the Arctic is Warming So Much Faster than the Rest of the World, the Conversation, June 25, 2020. Available at: https://theconversation.com/siberia-heatwave-why-the-arctic-is-warming-so-much-faster-than-the-rest-of-the-world-141455.

[9] P.K. Quinn, T.S. Bates, E. Baum, Short-lived pollutants in the Arctic: their climate impact and possible mitigation strategies, Atmos. Chem. Phy. 7 (2007) 15669−15692.

[10] Arctic temperatures highest in at least 44,000 years, Livescience (24 October 2013).

[11] G.H. Miller, S.J. Lehman, K.A. Refsnider, J.R. Southon, Y. Zhong, Unprecedented recent summer warmth in Arctic Canada, Geophys. Res. Lett. 40 (21) (2013) 5745−5751, https://doi.org/10.1002/2013GL057188.

[12] Seablind documentary mentioning that the burning of bunker fuel by ships contributes to black carbon deposits on snow and ice in the Arctic.

[13] The Race to Understand Black Carbon's Climate Impact, Climate Central, 2017.

[14] Luis A. and Roy N. Decadal Arctic Sea-Ice Variability and its Implication to Climate Change (Communicated) − Personal Communication.

[15] B. Cecilia, Polar Amplification, 2006. RealClimate.org.

[16] J.A. Screen, I. Simmonds, The central role of diminishing sea ice in recent Arctic temperature amplification, Nature 464 (7293) (2010) 1334−1337, https://doi.org/10.1038/nature09051.hdl:10871/10463. PMID 20428168.

[17] D. Ghatak, A. Frei, G. Gong, J. Stroeve, D. Robinson, On the emergence of an Arctic amplification signal in terrestrial Arctic snow extent, J. Geophys. Res. Atmos. 115 (D24) (2010), https://doi.org/10.1029/2010JD014007.

[18] Study Links 2015 Melting Greenland ice to Faster Arctic Warming, University of Georgia, 9 June 2016.

[19] M. Tedesco, T. Mote, X. Fettweis, E. Hanna, J. Jeyaratnam, J.F. Booth, R. Datta, K. Briggs, Arctic cut-off high drives the poleward shift of a new Greenland melting record, Nat. Commun. 7 (2016) 11723, https://doi.org/10.1038/ncomms11723.PMC4906163. PMID 27277547.

[20] E.L. Weiser, S.C. Brown, R.B. Lanctot, H. River Gates, K.F. Abraham, Effects of environmental conditions on reproductive effort and nest success of Arctic-breeding shorebirds, Ibis 160 (3) (2018) 608−623, https://doi.org/10.1111/ibi.12571.S2CID53514207.

[21] J. Yadav, A. Kumar, R. Mohan, Dramatic decline of Arctic sea ice linked to global warming, Nat. Hazards 103 (2) (2020) 2617−2621.

[22] Kumar A., Yadav J. and Mohan R., 2020. Global warming leading to alarming recession of the Arctic sea-ice cover: insights from remote sensing observations and model reanalysis, vol 6(7).

[23] IPCC AR4 Chapter 10 [1] Table 10.7.

[24] J.M. Gregory, P. Huybrechts, S.C. Raper, Climatology: threatened loss of the Greenland ice-sheet, Nature 428 (6983) (2004) 616, https://doi.org/10.1038/428616a.PMID15071587.

[25] Regional Sea Level Change. Intergovernmental Panel on Climate Change.

[26] https://www.ctvnews.ca/sci-tech/greenland-s-ice-sheet-has-melted-to-a-point-of-no-return-according-to-new-study-1.5065393?cache=juzexmjvhq%3FautoPlay%3Dtrue%3FautoPlay%3Dtrue.

[27] https://en.wikipedia.org/wiki/Arctic_resources_race#cite_note-53.

[28] https://en.wikipedia.org/wiki/Arctic_resources_race#cite_note-55.

[29] T. Edwards, The Arctic Heatwave: Here's What We Know, The Guardian, 2020. Retrieved 2 July 2020.

[30] Artic Report Card (2019) https://www.arctic.noaa.gov/Report-Card/Report-Card-2019.

[31] IPCC (2011) https://archive.ipcc.ch/pdf/special-reports/srren/SRREN_FD_SPM_final.pdf.

[32] AMAP (2013) https://www.amap.no/documents/doc/amap-assessment-2013-arctic-ocean-acidification/881.

[33] Arctic Monitoring and Assessment Programme (AMAP) 2018. Arctic Ocean Acidification, Tromsø.

[34] S.J. Hassol, Impacts of a Warming Arctic, reprinted ed., Cambridge University Press, Cambridge, UK, 2004.

[35] Archived from the Original (PDF) on 23 September 2013. Retrieved 5 November 2012.

[36] M. Nuttall, P.A. Forest, S.D. Mathiesen, Adaptation to Climate Change in the Arctic, University of the Arctic, 2008, pp. 1−5.

[37] M. Eckel, Russia: Tests Show Arctic Ridge is Ours" the Washington Post, Associated Press, 2007. Retrieved 21 September 2007.

[38] M. Humpert, A. Raspotnik, The future of shipping along the transpolar sea route, The Arctic Yearbook 1 (1) (2012) 281−307.

[39] As the Earth Warms, The Lure of the Arctic's Natural Resources Grows. https://www.popsci.com/science/article/2013-01/energy-development-arctic/.

Chapter 15

Environmental risk from exploitation of the Arctic

The Arctic is characterized by a harsh climate with extreme variation in light and temperature, short summers, extensive snow and ice cover in winter, and large areas of permafrost. Its terrain varies from high mountains to flat plain, wide tundra, and great expanses of sea, snow, and ice. The Arctic is under great threat from a multitude of environmental changes induced by human activities, most importantly through climate change, but also through pollution, industrial fishing, foreign species introduced to the area, nuclear waste, and petroleum activity. Air pollution affects Arctic environments in a different manner. While, toxic mercury, emitted into the atmosphere by coal-burning and industrial activity, is accumulating in the Arctic, threatening Arctic life, air pollution can also badly harm the important food source of lichen. The health risks from oil and gas extraction are not only through air pollution but also through contaminated drinking water sources with chemicals that lead to cancer, birth defects, and liver damage.

Climate change and the fossil fuel industry induced carbon emissions is an indirect threat to the Arctic ecosystem. On the contrary, diminishing Arctic sea ice poses a direct threat to the Arctic's biodiversity and eventually to the planet. Unprecedented Shipping traffic in the Arctic region increases the possibility of damage to its fragile marine environment. Any type of oil spill during drilling in the Arctic region becomes tricky owing to the vast size, remote location, and extreme weather conditions coupled with the complete lack of infrastructure.

It, therefore, can be understood that Arctic Ocean drilling is a gamble with catastrophic consequences for the people, wildlife, and the sensitive ecosystem of the region.

The accidental spill of oil and chemicals due to increased shipping activities in the Arctic is, although by no means the only concern, perhaps the greatest one [1]. Considering the added challenges of Arctic operations, the risk of accidents may increase in these waters. Presently, there are very few ways of convalescent spilled oil from ice-covered waters. These factors ought to be self-addressed to avoid severe ecological and economic consequences.

Other equally important concerns include the introduction of invasive marine species through ballast water discharges and the need for rigorous

The Arctic. https://doi.org/10.1016/B978-0-12-823735-9.00002-3

control of ships' garbage and waste products. Even carefully controlled shipping can cause unintended damage to wildlife if the shipping routes run through areas of critical environmental concern, such as whale foraging zones or migration corridors. Ship strikes are already one of the most significant threats to the survival of the northern right whale. Indeed, most of the environmental dangers posed by shipping activities are not due to explicit accidents like the accidental large-scale oil spills but rather to routine shipping operations [2]. Increased shipping in the Arctic must be conducted in an environmentally sound fashion.

As mentioned earlier that Oil and gas drilling is a dirty business having serious consequences for our wildlands and communities. Drilling projects operate around the clock, disrupting wildlife, animal water sources, human health, and many other aspects. Dangerous emissions contribute to climate change, and longer wildfire seasons are a consequence of the planet's rising temperatures. Drilling induced pollution ruins pristine landscapes, and adversely impacting the vegetation. It may lead to increasing erosion, which in turn initiate landslides and flooding. It therefore, disturb the land's ground surface, and seriously fragment unspoiled wildlife habitats. It has been observed that the Light pollution is also impacting wildlife and wilderness. Therefore, responding to oil spills due to drilling in the Arctic Ocean is extremely dangerous as our ability to respond to emergencies and oil spills is severely limited and thus poses the underpinning motivation for precautionary and prohibited actions in the larger interest of the Antarctic environment and its conservation.

1. Environmental issue of Arctic

The Arctic environment is responding very sensitively to global warming, and the Arctic Ocean sea-ice is decreasing at a pace exceeding scientific predictions. Over the past 35 years, the Arctic sea ice extent in the summer has declined by nearly two-thirds. If effective mitigation measures are not taken, and if global warming continues to accelerate at the maximum pace, a nearly ice-free Arctic Ocean in the summer by the mid-century is likely. Although the mechanisms of environmental modification in the Arctic are still not sufficiently understood, the impact of global warming is amplified to a bigger extent in the Arctic than in any other regions on the planet. There is a risk that rapid change in the Arctic environment will have a drastic and irreversible impact on the foundations of the lives of indigenous peoples and others who live in such a harsh environment, and on the ecosystem under the vulnerable environment in the Arctic (Fig. 15.1). Therefore, the international society needs to act in a responsible manner. There are also concerns that environmental changes in the Arctic will accelerate global warming, leading to global sea-level rise, increasing the frequency of extreme weather events, and adversely affecting ecosystems [3].

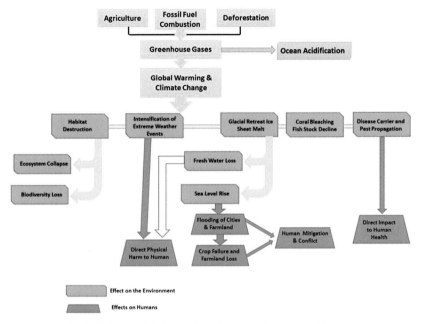

FIGURE 15.1   Global warming effect on environment and human.

At the same time, the decreasing quantity of ocean ice has dilated the passable space, enabling the gap of shipping lanes within the ocean and different new economic uses. Amid increasing interest in economic activities in the Arctic, including the development of mineral and marine living resources and utilization of the Arctic Sea Route, discussions are underway at the Arctic Council, the International Maritime Organization, and other forums regarding economic activities that can be carried out in a sustainable manner while preserving the vulnerable and low resilient Arctic environment as well as international rule-making. Some Arctic states, with a read toward securing their national interests and protective their territories, became active within the space of national defense [4].

It means, changes within the Arctic have political, economic, and social effects. Such impacts are not confined to the Arctic but tend to transgress globally. The ensuing opportunities and problems are attracting the eye of the world community including the Arctic and non-Arctic states. Rapid environmental changes in the Arctic, while the changes are increased and amplified by global environmental changes, should be regarded as global issues beyond regional issues because it is possible that the changes in the Arctic will affect globally including global warming (Fig. 14.4A). Although the major factor of climate change is global warming caused by increased emissions of greenhouse gases, the accelerated warming in the Arctic is caused by the interaction of several factors in a complex process, including the atmosphere, sea currents, and especially the decrease of sea ice in the Arctic Ocean. It permits additional open water to soak up heat from the rays of rising temperature in the Arctic due to Global warming. There are issues that amendment within the Arctic setting might increase the frequency of utmost weather events in Japan and alternative mid- and high-latitude states. It is projected that warming within the Arctic can seemingly continue, even within the absence of dilated economic activity. Furthermore, it is identified that developmental activities can end in pollution of the air and water, like unseaworthy and discharge of pollutants from ships into the ocean [5].

A unique Arctic ecosystem requires a careful and responsible attitude, so the economic development of the region actualizes the problem of compliance with environmental protection requirements and environmental protection both from Russia and from its foreign partners. There remains a risk of radioactive contamination of the Arctic, and the negative impact on polar ecosystems also provides atmospheric streams and currents that bring North pollutants from Western Europe. In addition, the problem of preservation of the Arctic biosphere is also pressing against the activities of foreign companies in the far North. The state should carry out the assessment of the impact on the environment and ecological examination, to develop mechanisms of financial responsibility for domestic and foreign companies in the event of an accident, as well as support the development of new competitive and environmentally friendly technologies. One constant throughout the Arctic region is the hostile climate. Record-setting cold, ice-covered waters, quickly rising storms, and high winds outline the region. The warming trends of the High North are about double the rate of global warming trends and such a magnitude that the pace of sea ice decline and surface ocean warming is unprecedented. This warming is conducive to a decline in ice coverage. The warming trends are forecasted to continue at a progressively speedy rate due to increased reflective power, creating Arctic weather additional unpredictable because of the probability of fog, storms, and even ice floes rise in future years [6].

The vegetation, animals, and residents of the Arctic region getting into trouble due to many reasons. These problems are becoming grave day by day for them. The main cause of these problems is environmental change due to the intentional and unintentional human activities. The strange fact is that most of the direct or indirect impact of these activities must be borne by the natural resources and habitats of the Arctic region. They will have to bear the impact in the future as well. So, the measures to protect the Arctic must be started from other regions. It is a universal fact that, due to human activities, excessive secretion of greenhouse gases is a major factor for environmental change whether the increase is due to the industrial revolution or excessive deforestation or increase in agricultural products. So, to stop or to lessen the unwanted changes, we need to curb the emission of greenhouse gases. The efforts to decrease the level of emission is known as Mitigation among scientists.

Though by a reduction in emission of greenhouse gases, the adverse effects of environmental change will be reduced, but due to the man-made errors during the previous centuries and especially in the beginning of the century, the greenhouse gases have been accumulated in the environment. These gases can persist for a very long period without a chemical breakdown in the environment even for hundreds of years. In this way, the greenhouse gases accumulated in the environment in different ways will have adverse effects in the future. Therefore, to minimize the adverse effects of greenhouse gases, we need to adapt ourselves in such a way that these gases do not harm us. This process is called adaptation. Let us discuss about the mitigation first. The Arctic has a unique changing environment despite a large number of studies carried out in the Arctic region, and it is not well understood. There is a need for sound scientific and socio-economic information as the Arctic environmental research, monitoring, and vulnerability assessments are considered top priorities. These priorities include sea ice and glaciers, thawing permafrost and coastal erosion, as well as the pollutants from the Arctic contaminating the region. As the temperature rises, contaminants locked in the ice and soils will be released into the air, water, and land. Due to an increase in human activity in the Arctic region, there would be an increase in contaminants concentration [7]. The Arctic region is a vast area whose economic potential, through climate change and advanced technology, is becoming accessible for the first time in history. This potential includes hydrocarbon resources as well as shipping lines, fishing rights, and metal deposits. Developing the Arctic will require massive investment particularly in transportation, extraction, and governance infrastructure. Yet elsewhere establishing substructure, Arctic states must also consider the various environmental and diplomatic risks associated with such development. While climate change is helping to fuel Russia's moves in the Arctic, Russia's development of the region will continue contributing to rising temperatures. This, and the fact that Arctic development presents specific environmental risks, makes Russia's activities in the region of relevance within international environmental discussions. The greatest environmental threat to the region stems from the extraction and transportation of fossil fuels. While oil and chemical spills are likely to occur in any region containing oil and gas reserves, when such spills occur in the Arctic, they are particularly dangerous incidents [8].

For example, because the Arctic drilling season is typically limited to a few months during the summer, companies have a limited timeframe to cap leaking oil/gas wells or successfully drill relief wells before winter sea ice returns. Oil spills are also difficult to clean up and contain, as the oil will become trapped under large blocks of ice and travel large distances under ice flows. Extreme weather conditions, short days, lack of search, and rescue stations also amplify the danger of such events. Arctic shipping also poses environmental threats. For example, black carbon (soot) is a common pollutant produced by marine vessels through the incomplete oxidation of diesel fuel. When in the air, black carbon particles absorb sunlight and generate heat in the atmosphere, affecting cloud formation and rain patterns. When covering snow and ice, the particles absorb the sun's radiation, generating heat and speed up the melting process. In 2004, 609 tons of black carbon was released into the Arctic region. Shipping other raw materials, such as nickel, also poses major risks, as raw nickel is a known carcinogen. Coastal waters where most of the shipping takes place also often contain high levels of biodiversity and many Arctic shipping routes coincide with spring migration paths used by many marine mammals, including bowhead, beluga, narwhal, and the walrus, to reach their summer feeding grounds. When longer shipping seasons coincide with short migration and feeding seasons, such disruption may result in many species' inability to survive the winter [8].

2. Mitigation

Most of the emission of greenhouse gases is due to the transportation and power production industries especially due to the burning of fossil fuel. It has been opined that the contribution of transportation is 33% and power production industries are 25.9% in the emission of greenhouse gases. Approximately 45% of the land of the earth is being used for agriculture. About 13.5% part of the total amount of greenhouse gases is emitted due to agriculture. To fulfill the needs like food, accommodation, and livelihood of the increasing population, excessive deforestation is going on for the past few decades. Due to this reason, the carbon dioxide being used by the plants and trees is also reducing. Therefore, it is increasing in the environment. It is contributing to the emission of greenhouse gases by 17.4%. It is a significant fact that according to the scientists during 2003, approximately 395 crore hectare of earth land was covered by forests but despite the best efforts to protect the forests the rate of deforestation is 7.2% every year [9].

For the past decade, there is considerable awareness among the masses about environmental changes. Crossing the boundaries of scientific labs and debates by political leaders, this subject has come to the masses now. Therefore, most of the countries have become concerned about the adverse impacts and they are thinking seriously to curb and mitigate the environmental changes at the national and international level. The United Nations have taken active and concrete steps. Vast Expert Committees comprising members from each country have been constituted and every country took part in the deliberations of the committees. The last conference in this sequence was organized in Mexico in 2010. Heads and diplomats from 195 countries participated in the Copenhagen Conference in 2009. The country heads delivered long speeches during the conferences, discussions were held for hours and crocodile tears were shed about the future of the human. Many protocols were prepared. Solid steps were suggested regarding the mitigation of greenhouse gases and quantities were determined for each country in this direction for a certain period.

Many countries, except those developed countries that emit more quantity of greenhouse gases, signed the protocol and they started implementing those procedures, but the implementation was hindered by the political selfish motives, commercial benefits, and industrial development by rapid consumption of nonrenewable resources (like coal and fossil fuel) and being a part of the growth race with the developed nations. These hindrances are still there. Therefore, the mitigation drive could be implemented successfully, and the Arctic is suffering due to the adverse effects of environmental changes. Usually, to minimize the amount of carbon dioxide in the environment, methods like dense forestation and dumping of some amount of carbon dioxide in deep pits of the earth are suggested. The capacity of the ocean to dissolve the extra amount of carbon dioxide in the environment cannot cope with the increasing level of carbon dioxide and the forestation and dumping technique is not proving successful.

Therefore, to fight with ill effects of climatic change under the circumstances, people are focusing on another technique called adaptation.

3. Adaptation

There is an old popular saying, "if you are not able to change, develop the habit to adjust with the conditions." This saying is true about the climatic change like many other circumstances of life. We have to produce paddy and other cereals in large amounts to fulfill the requirements of food-stuffs of the rapidly increasing human population. In this way, to fulfill the demand for milk and meat, animal husbandry should be practiced. Mostly, carbon dioxide is considered responsible for Global Warming because of its largest quantity of greenhouse gases. Transportation and power industries generate the major quantity of carbon dioxide because fossil fuel is used for them. Though many engines have been invented, which function by Solar Energy and Electricity, but their quantity is not enough. We need to make them functional and popular. At the same time, we need to promote the use of public transport so that people do not use personal cars on a large scale. The countries that have large reserves of coal and petroleum generally use them as fuel. Even the big industrialist without paying any attention to the threat of global warming and give arguments for the use of petroleum and coal, but to avert the threat of environmental changes and global warming we need to adopt nonconventional methods of energy like solar energy, wind energy, and other natural resources.

Though the electricity is being produced by these renewable sources, its use is negligible. We need to use these sources on a large scale so that the needs of industrial metro cities could be met, and economic and simple techniques can be developed [10].

Today developed and prosperous countries are emitting greenhouse gases at a large scale and the United States is a burning example for that. The population of this country is just 4% of the total population of the world; still, it is generating 25% of the total quantity of greenhouse gases in the world. Despite that, a powerful lobby is also there which is opposing the abetment-related protocols of greenhouse gases directly or indirectly. Even technically developed countries such as the United States can invent economic and modern techniques for large-scale production. The environmental changes are adversely affecting every aspect of human life, economic, social, political, and industrial sectors. Therefore, we need to keep sustainable development in mind while providing solutions to them. The emphasis should be on understanding that mitigation and adaptation are not options, but they are complimentary to each other. Both techniques need to be used simultaneously. Although both the techniques are very expensive, in various surveys it was found that for applying mitigation techniques, the country may have to spend 2% of their Gross Domestic Product. This amount is so big and because of that most of the countries are applying this technique at a very low pace [11].

As compared to mitigation, the adaptation technique is cheaper. Economically strong and technically developed countries may use both the techniques in a comparatively better way. Therefore, the side effects of climate change in these countries are lesser.

The adverse effects of environmental changes are to be borne by the poor and helpless ones rather than the prosperous and rich people. Most of the adverse effects are on the tribal people who have lagged in technological development. The total population of these tribes is approximately 37 crores. It includes a 40 lakhs tribal population of the Arctic Region. In this connection, a strange fact is that their lifestyle is such that emission of greenhouse gases is minimal, but they are being affected by these gases emitted by other regions. It will not be an exaggeration that the Arctic is the Barometer of the world's climate change and the tribesmen are its mercury.

The Arctic is in those countries that are considered the most developed and growing countries. These countries are Canada, the United States, Russia, Finland, Sweden, Norway, and Denmark. Although these countries are technologically developed, yet their record in minimizing the greenhouse gasses has been very poor. Sweden is the best country in this aspect, yet its record cannot be considered as very good. In other words, if we follow the standards of Sweden for the emission of carbon dioxide, its quantity in the atmosphere will never decrease. In this aspect, if Sweden's record is good, Norway and Finland's record is below average. Canada, the United States, and Russia are considered in the 10th highest pollution generating countries. Even with such a worst record, these countries are not taking any steps in minimizing the emission of greenhouse gases. They provide financial help to their industrialists to use solar and wind energy and also provide funds to develop efficient automobile technologies and promote people to save energy. The country also provides funds to other developing countries to develop techniques to save energy and reduce the emission of greenhouse gases.

The total population of tribes in the Arctic Region is 10% of the total population. In the Arctic regions of Canada, it is at most 50% and in Greenland, they are in majority. The rest of the population is of migrated people. They came there for fishing, mining, and in the search of other work. Natives of the Arctic mainly depend on hunting for their food and follow the custom of their ancestors. These are those people who produce the minimum amount of greenhouse gases but are affected the most.

In this context, the most unbelievable fact is that the representation of these tribes in the Committees of the United Nations and other international summits on environmental changes is negligible. Their opinion is not taken on the environmental issues of other countries as well. Therefore, in 2002, during the Eighth summit organized in New Delhi by United Nations Framework Convention on Climate Change, the representatives of these tribes had made a strong appeal for their equal contribution in the summits of climate change. But till now, the appeal has not been considered. The people and animals that live in the Arctic depend on its unique ecosystem to survive. For them climate change is not a debate, it is a daily reality as with the region warming twice as fast as the rest of the world, Arctic ice is melting even faster, due to the fact that the ocean absorbs the heat.

References

[1] A. Chircop, International Arctic shipping: towards strategic scaling-up of marine environment protection, in: M.H. Nordquist, J.N. Moore, T.H. Heider (Eds.), Changes in the Arctic Environment and the Law of the Sea, 2010, pp. 177−202.

[2] A.K.Y. Ng, S. Su, The environmental impacts of pollutants generated by routine shipping operations on ports, Ocean Coast Manag. 53 (56) (2010) 301−311.

[3] T. Williams, The Arctic. Organisations Involved in Circumpolar Cooperation, Publication No- 2008-15 E, Parliamentary Information and Research Service, Library of Parliament, Ottawa, 2012.

[4] S.L. Donald, J.W. Fiona, Climate Change Impacts and Adaptation: A Canadian Perspective, Natural Resources Canada, Ottawa, 2004.

[5] R. Kerr, Is battered Arctic sea ice down for the count? Science 318 (2007) 5847.

[6] J.C. Comiso, D.K. Hall, Climate trends in the Arctic as observed from space, Wiley Interdiscip. Rev.: Clim. Chan. 5 (3) (2014) 389−409.

[7] https://en.wikipedia.org/wiki/Arctic_policy_of_the_United_States#cite_note-Bush-60).

[8] https://geohistory.today/russia-arctic-development-power/). http://csef.ru/en/politica-i-geo-politica/501/problemy-militarizaczii-arktiki-v-sovremennyj-period-6440, Publication date: 14-11-2015).

[9] J.E. Overland, M. Wang, J.E. Walsh, J.C. Stroeve, Future Arctic climate changes: adaptation and mitigation time scales, Earth's Future 2 (2) (2014) 68–74.

[10] F. Chris, D.P. Terry, in: D.S. Lemmen, J. Lacroix, F.J. Warren (Eds.), "Northern Canada," in from Impacts to Adaptation: Canada in a Changing Climate 2007, Nat. Res. Canada, Ottawa, 2008, pp. 57–118, 2008.

[11] J.E. Overland, M. Wang, N.A. Bond, J.E. Walsh, V.M. Kattsov, W.L. Chapman, Considerations in the selection of global climate models for regional climate projections: the Arctic as a case study, J. Clim. 24 (2011) 1583–1597.

Chapter 16

Militarization of the Arctic

The Arctic has long been considered as the world's "last frontier," the "last white dot on the map." The Arctic region is located around the North Pole and is surrounded by the landmasses of five Arctic countries, namely Canada, Denmark (via Greenland), Russia, Norway, and the United States. Currently, no country owns the North Pole. It sits in international waters. The closest land is Canadian territory Nunavut, followed by Greenland (part of the Kingdom of Denmark). However, Russia, Denmark, and Canada have staked claims to the mountainous Lomonosov Ridge that runs under the pole. Today, the Arctic is routinely described as an emerging frontier and many polar nations along with a few that have no Arctic borders are angling for access to the region's rich stores of fish, gas, oil, and other mineral resources. The militarization of the Arctic was dictated by geography and shaped by technology, which is believed to be the sole thrust to transform the Arctic from a "zero military zone" before World War II, to a military flank in the 1950−70 period, and a military front in the 1980s. The need to project national influence and sustain claims over the region's sea-lanes and natural resources justifies the rising military presence in the Arctic.

The Arctic has long been defined by the harsh climate and ice-choked waters owing to its isolated and remote area. The Arctic is speculated to hold oil reserves of up to 13% of the global total of undiscovered oil, up to 30% of natural gas, and other precious metals. These resources were previously inaccessible due to the layers of thick ice that cover and surround the Arctic region. But now, the ice is melting rapidly in the region because of global warming, making this isolated and inaccessible region more viable for economical exploitations.

In the recent past, the melting of the polar ice caps is not only opening novel opportunities for commercial activities but also a more open and hospitable Arctic has also led to increased territorial claims and military presence by the stakeholding countries of the Arctic region (Fig. 16.1).

Nevertheless, due to the Arctic Council, the establishment in 1996 with the sole objective of providing an integral means for cooperation, coordination, and interaction among Arctic states, the prospect of a conflict in the Arctic remains implausible.

As mentioned earlier, the Arctic holds billions of barrels of oil and gas and other natural resources beneath it. With Global warming, the ice coverage of

The Arctic. https://doi.org/10.1016/B978-0-12-823735-9.00008-4

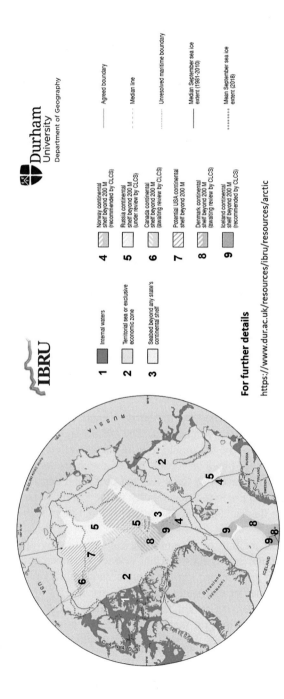

FIGURE 16.1 Territorial claims over the Arctic region by various countries. *Source: https://www.dur.ac.uk/resources/ibru/resources/ArcticMapsMay2020/ Arcticcontinentalshelf-compressedpp.pdf.*

the Arctic is shrinking considerably making the once inaccessible Arctic frontier accessible. The Arctic nations are aggressively making competing claims for their sovereignty considering the discovery of huge resources in the Arctic.

The Arctic resources are becoming exploitable and as one of the last places on Earth to be exploited for minerals, it is likely to be a valuable area. The difficulty arises because there is a large handful of nations that have legitimate claims to parts of the area. Their claims are sometimes conflicting.

An ice-free Arctic Ocean as prophesied by climate change experts (Fig. 16.2) may emerge as a game changer of the existing geopolitics through the most sweeping transformation of its dimensions including commercial, constabulary, military components, and economic opportunities added to this enterprise. As a well-known fact that countries fronting on polar waters, the United States, Canada, Denmark, Finland, Iceland, Norway, Russia, and

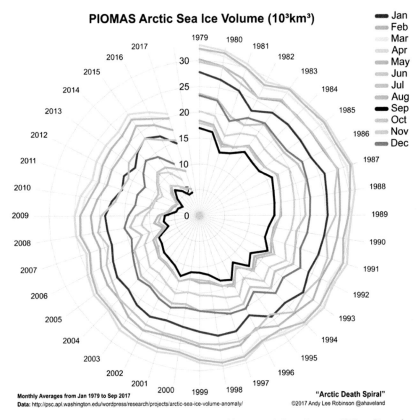

FIGURE 16.2   Changes in the sea ice extent (monthly average) from January 1979 to December 2017 period. *Source: https://upload.wikimedia.org/wikipedia/commons/3/33/Arctic-death-spiral.png.*

Sweden comprise the intergovernmental Arctic Council, are all currently jostling for ownership of the region's frozen seas, and will enjoy exclusive rights to fish and tap undersea resources in hundreds of thousands of square miles of water off their shores mainly because of their geopolitical advantages, and thus emerge out as the main "players" in the strategic games.

Military capabilities within the Arctic have steadily exaggerated over the past 10 years by way of setting up new airfields, naval command bases for the region, increasing the icebreaker fleet. Within the context of the growing interest of world powers within the Arctic, there is a dangerous tendency of the militarization of the region. This can be clearly apparent within the prominent military presence and activity of states within the Arctic, modernization of the military and infrastructure within the way north, and intensive use of the military to guard its economic interests. Several States these days pursue a policy of fixing the legal regime of the Arctic. The main target is going to be mining and therefore the development of transit routes. With the introduction of the system of governance for the Arctic providing access to the Arctic region, for new players, the initial Arctic States might lose a part of their influence.

It is obvious that nations custodian of waterfront property in the Arctic, which is rich in strategic resources, will take all security measures to safeguard their territorial seas and exclusive economic zones during ice-free intervals through the deployment of Coast guards and military personnel. Competitions for interests and rights in the Arctic have resulted in an escalating militarization of the region.

1. Russia

In recent years, Russia has increased Arctic military activities such as drills, opening or reopening military bases, construction of icebreakers, and establishment of advanced radar stations to enhance its control of the region. Further, in 2017, Russia unveiled an Arctic Trefoil military base on Alexandra Land Island in the northeast Barents Sea. It is Russia's northernmost military outpost and the second of its kind, following the completed Northern Clover on Kotelny. The main objectives of Russia in its Arctic policy are to utilize its natural resources, protect its ecosystems, use the seas as a transportation system in Russia's interests, and ensure that it remains a zone of peace and cooperation. Following the Russo-Georgian War in 2008, Russia has maintained a large presence in the disputed regions of Abkhazia and South Ossetia. The Russian 4th Military Base is in South Ossetia and hosts around 3500 personnel. Russia currently maintains a military presence in the Arctic and has plans to improve it, as well as strengthen the Border Guard/Coast Guard presence there. Using the Arctic for economic gain has been done by Russia for centuries for shipping and fishing. Russia's consideration for reopening a major northern naval base and resuming regular naval patrols has generated a debate over the militarization of the Arctic. The Russian decision to rebuild a naval facility in the Arctic is seen as preparedness toward a futuristic scenario when the northern ice-cap melts and critical sea-routes become navigable, Arctic nations will not be able to resist the impulse of militarizing the region and thus increasingly assertive territorial postures will be adopted by regional stakeholders for gradual pontification of security-driven planning and actions. It is important for Russia to assert control over the operation of the Northern Sea Route (NSR).

Ever since (2007), a mini-sub planted a Russian titanium flag at the base of the North Pole confirming a strategic presence in the Arctic, Russia has emerged out to be the most militarily active Arctic state. Further addition of new Borey class submarines, on the Barents Sea coast to its under-sea patrol program, will notably reflect Russia's power-projection initiatives. Russia's first-ever amphibious landing on the Arctic archipelago of the New Siberian Islands is yet another testimony to its intentions.

Recently submitted claim to the United Nations to extend Russia's 200-mile Exclusive Economic Zone by another 150 miles or 1.2 million square kilometers is arguably contested by other Arctic stakeholders. Russia is, however, not the only country with plans to securitize the region. Russia is continuing to show signs of hardening its Arctic strategies by introducing certain new regulations on the use of the NSR by foreign vessels. It would be obligatory on such vessels to share the information on ships and their cargo with Russian authorities.

Russia has a simple approach toward the Arctic region to protect Russia's political and economic interests in the Arctic. Moreover, Russia even has formally set the goal of deploying a combined-arms force within the Arctic region as well as military, border, and coast guard units by 2020, to guard its political and economic interests within the Arctic and boost Russia's military security.

However, the potential for conflict of interest like Russian—Canadian claims to the ocean bottom at the pole space may not be undermined as it may probably be converted into a security issue. The Russian military has been suggested to "devote special attention to deploying infrastructure and military units within the Arctic." Russia is out and away from the country with the most important military capabilities within the Arctic. The Northern Fleet is the fleet of the Russian Navy in the Arctic. Its headquarters and main base are located in Severomorsk, Murmansk Oblast.

2. Canada

Canadian initiatives to strengthen its Arctic sovereignty claims and reinforce its northern military presence recommended permanent military presence in remote locations in the Arctic North. The annual *Operation Nanook* in the country's north is an exercise aimed to foster Canadian sovereignty, which is in addition to *Operation Nunalivut* in the High Arctic and *Operation Nunakput* in the western Arctic. Canada strongly opines that Arctic Council must focus to seize the economic opportunities arising from the melting of the northern polar ice cap by bringing appropriate changes in its mandate.

While control over the Northern Sea Route has received much attention nonetheless, the gradual opening of the Northwest passage could also pose an equal challenge. It is significant to note that Canada's sovereignty claims over part of the Beaufort Sea, which is also holding the Northwest passage is under active contest by many stakeholder countries including the United States, to have the legal status of "international waters."

3. Denmark

Denmark and Canada have mutually agreed for collaborative efforts in the entire gamut of consultations, information exchanges, visits, and military exercises in the Arctic region. Participation of Denmark in the Canadian's Operation Nunalivut exercise in the High Arctic further demonstrates their solidarity with Canada. Denmark has highlighted the growing geopolitics importance of the region through its continuing management of Kalaallit Nunaat and therefore the Faroe Islands, together with its commitment to produce a separate Arctic command associate as an Arctic Response Force.

4. Norway

Norway conducted one of the "Exercise Cold Response," which is apparently the largest Showcasing of Norwegian intentions for military flank in the Arctic region, as it fears that Russia may take over Barents Sea despite having an important pact with Russia in 2010. Norway has substantial fisheries and fossil fuel interests in its Arctic territories and Exclusive Economic Zone (EEZ). Its military strategy has historically been mounted on Russia; however, within the last decade, it has shifted to concentrate on conflicts of interest within the Arctic space. The bulk of its military forces and installations are settled north of the Arctic Circle, and therefore the navy has been upgraded with a crucial concentrate on higher operational capability in Arctic waters.

5. The United States

The United States is one of the eight nations surrounding the Arctic along with Canada, Denmark (via Greenland), Finland, Iceland, Norway, Russia, and Sweden that are all currently jostling for ownership of the region's frozen seas. All land, internal waters, territorial seas, and EEZs in the Arctic are under the jurisdiction of one of these eight Arctic coastal states. The United States has signed, but not yet ratified the UNCLOS. International law regulates this area as with other portions of the Earth.

The Thule Air Base is a military base in Greenland belonging to the United States. It is, in fact, the United States' northernmost military base in the world, located just 1524 km from the North Pole and 1207 km north of the Arctic Circle. Although US nuclear submarines have operated under the ice for decades, nevertheless its surface navy is ill-equipped for the Arctic. The United States understands the need for a strategic Arctic port in Alaska and navigational operations in northern shipping lanes by its Navy, to help build up the capacity to conduct emergency operations.

No doubt, the US warship sailing near Russia's vast northern border would be viewed as a matter of concern for Russia mainly related to the geopolitical challenges caused by global warming and associated opportunities. The United States advocates to project military strength in the Arctic without any intention to be unduly provocative.

The United States felt it essential to strengthen its Arctic strategies. Accordingly, the Russian's rules have been deprecated with a hope to conduct navigation operations freely' in the Arctic. The United States reiterates the security challenges of China and Russia in the region. Using its economic power, China may seek to influence Arctic governance. Enhanced awareness of Arctic challenges and increased operations in the region must be the top agenda of the United States. The United States and Russia have direct access to the Arctic region, in contrast to China.

The US DoD's Arctic Strategy (2016) supports the enhancement of the ability of forces of the United States to defend the homeland and exercise sovereignty by strengthening deterrence at home and abroad through alliances and partnerships. In the Arctic, it also preserves the freedom of the seas and involves international, public, and private partners to help improve domain awareness in the Arctic region. The policies also envisage the development of Arctic infrastructure and capabilities appropriately in tandem with changing conditions and requirements. It is imperative to support civil authorities, other departments, agencies, and nations toward environmental security. It further advocates the promotion of international regional cooperation within the ambit of governing laws over the Arctic region.

6. China

China's scientific interests in the Arctic region are viewed with cautions as its Arctic military strategies, including submarine deployments, are well known. China's intentions to expand economic interests through mining and infrastructure development in Greenland and to build a nuclear-powered icebreaker are perceived to be of serious concern. The ambitious plan of China to develop an "Ice Silk Road" and spearheading new organizations in the Arctic region is taken as a signal that China is not going to accept being marginalized in the region. To further strengthen its control over Arctic affairs, China has already established strong ties with other Arctic states. Similar to the United States and

Russia, China does not have direct access to the Arctic therefore to compensate that China has been deploying its wealth and diplomatic resources to secure its interests within the Arctic region [1].

Arctic nations have in recent years disclosed plans to upgrade military capabilities, providing operations within the harsh Arctic region (Fig. 16.3). In the year 2007, Russia symbolically staked its claim to the waters around the pole by planting a Russian flag on the Davy Jones. Similarly, the Canadian government declared plans to buy up to 18 military vessels capable of defensive Canadian sovereignty within the Arctic region. Whereas, the Norwegian national strategy revealed that it staked its claims within the Soria Moria Declaration to the northern Norway and thus implicitly the Svalbard dry land (Figs. 16.4 and 16.5) depicts an average preparedness of countries that

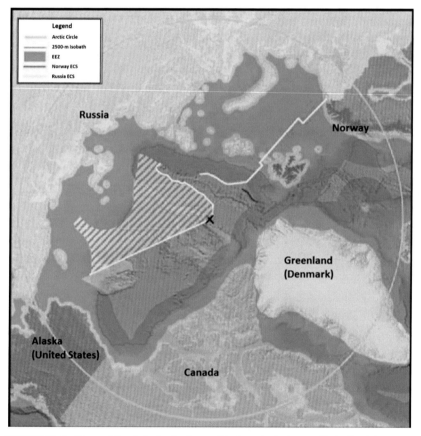

FIGURE 16.3 Claim of Russia in the Arctic sea. The large hashed area reflects Russia's current extended continental shelf claim. *Source: https://upload.wikimedia.org/wikipedia/commons/2/20/ Russian_Arctic_claim.PNG.*

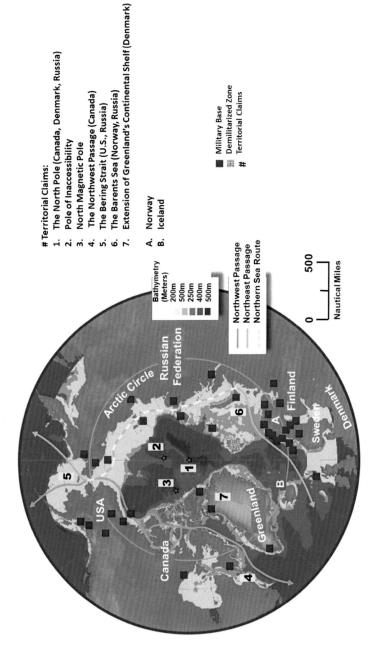

FIGURE 16.4 Militarization in Arctic (modified). *Source: https://en.wikipedia.org/wiki/Arctic_shipping_routes.*

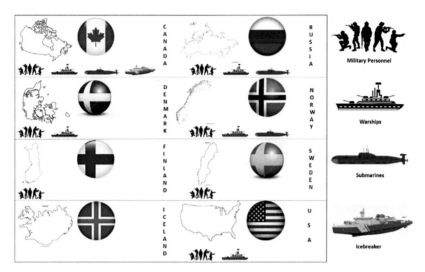

FIGURE 16.5 Deployment of military in Arctic by Arctic Nations.

constitute the intergovernmental Arctic Council fronting on polar waters namely the United States, Canada, Denmark, Finland, Iceland, Norway, Russia, and Sweden.

Warming has created more and more potential to use the substantial deposits of natural resources within the Arctic. Growing attention to the present potential chance for gain has alerted Arctic nations Russia, Norway, Denmark, Canada, and therefore the United States a lot to be watchful in defending their northern territories.

The Arctic is now facing a potential regional security threat. Not only the major Arctic strategic players, like Russia and the United States, but the non-Arctic states such as China are also attempting to enter the region with enforced regulations and procedures.

Reference

[1] http://www.lai.lv/viedokli/arctic-a-mirror-of-great-powers-geopolitical-interests-305.

Index

Note: 'Page numbers followed by "b" indicate boxes, those followed by "f" indicate figures and those followed by "t" indicate tables.'